RAND AUSTRALIA and
NATIONAL SECURITY RESEARCH DIVISION

T0302890

CONTESTABILITY FRAMEWORKS

AN INTERNATIONAL HORIZON SCAN

Cynthia R. Cook | Emma Westerman | Megan McKernan
Badreddine Ahtchi | Gordon T. Lee | Jenny Oberholtzer
Douglas Shontz | Jerry M. Sollinger

Prepared for the Australian Department of Defence

For more information on this publication, visit www.rand.org/t/RR1372

Library of Congress Cataloging-in-Publication Data is available for this publication.
ISBN: 978-0-8330-9278-6

Published by the RAND Corporation, Santa Monica, Calif.
© Copyright 2016 RAND Corporation
RAND® is a registered trademark.

Cover: Midshipman Rachel Bolitho looks out from the bridge wing of HMAS Canberra
(photo by LSIS Helen Frank/Royal Australian Navy).

Support RAND
Make a tax-deductible charitable contribution at
www.rand.org/giving/contribute

www.rand.org

Preface

The Australian Department of Defence (ADoD) is undergoing a fundamental restructure, one aspect of which aims to ensure that it has a robust military capability acquisition process. A key component of this restructuring is the establishment of an internal contestability capacity to review and challenge ADoD requirements, acquisition, and budget decisions internally before they are passed to other elements in the government. The role of this contestability function is to help ensure that the requirements and the resultant capabilities delivered to the Australian Defence Force are aligned with articulated strategy and agreed-upon resources. To help develop this capacity, the ADoD, in September 2015, engaged the RAND Corporation to survey and assess international practices in this arena.

This report details our findings. It describes key contestability functions and the primary aspects of those functions, as described in the literature. The report also provides the results of case studies of public and private organisations that have contestability functions.

This research was conducted within the Acquisition and Technology Policy Center of the RAND National Security Research Division (NSRD). NSRD conducts research and analysis on defence and national security topics for the U.S. and allied defence, foreign policy, homeland security, and intelligence communities and foundations and other nongovernmental organisations that support defence and national security analysis.

For more information on the Acquisition and Technology Policy Center, see www.rand.org/nsrd/ndri/centers/atp or contact the director (contact information is provided on the web page).

Contents

Figures and Tables

Figures

Tables

Summary

Next to the conduct of war, one of the most important functions of a nation's defence organisation is to invest its military with capabilities that deal with future threats. This function often requires that defence organisations make major expenditures, which attract both public and political scrutiny to ensure that the public's money is spent wisely. But large organisations of any sort face strategic choices that may fundamentally affect their future success and perhaps even survival. To ensure the most-effective decisionmaking, large organisations often have checks and balances as they make decisions about resource allocation. These checks and balances can take a variety of forms as organisations review decisions or even focus on strengthening decisions as they are in the process of being made. Checks and balances usually are built into the organisation's bureaucracy and are adjusted from time to time to take into account changes in circumstances, resourcing environments, or organisational culture. At other times, the processes by which these checks and balances are made become ossified and need a major shake-up.

The recently completed report *First Principles Review, Creating One Defence* criticised the current processes through which the Australian Department of Defence (ADoD) develops military capabilities.[1] Among the report's recommendations was a call for the establishment of an arm's length contestability[2] function in the ADoD to provide

---

[1] ADoD, *First Principles Review, Creating One Defence*, Canberra, April 2015.

[2] In this report, we define *contestability* as processes that organisations put in place that enable stakeholders to challenge, engage, and respond to decisions about the allocation of resources, including major acquisition efforts. It is a governance function to ensure that

"assurance to the Secretary that the capability needs and requirements are aligned with strategy and resources and can be delivered."[3] This would be the instantiation of a new process to review decisions about major resource allocation within the ADoD.

As it strives to implement this recommendation, the ADoD engaged the RAND Corporation, in September 2015, to identify practices, procedures, and elements that have been successfully implemented by foreign departments of defence and related agencies and by large commercial organisations in Australia and abroad. This report documents the results of RAND's international scan. It describes the relevant literature and case studies, along with the results of discussions that RAND conducted with representatives of U.S., European, and global organisations about the extent and nature of contestability functions at their institutions, conducted in September and October of 2015. The goal of the report is to draw out common themes and successful practices that may help the ADoD implement the *First Principles Review*'s recommendation.

To conduct this analysis, RAND employed a qualitative research approach—involving a literature review, examination of models of existing contestability functions, and discussions with experts on how contestability operates in different contexts. This resulted in an international scan of public- and private-sector approaches to contestability.

Reviewing Contestability Literature

We reviewed an array of organisational and management literature related to contestability.[4] It should be noted that there is no extensive

---

capability needs and requirements not only align with strategy and resources but also are delivered effectively.

[3]   ADoD, 2015, p. 38.

[4]   While broad, our review was not intended to be an exhaustive and systematic overview of the entire range of relevant work (which would cover management theory, evaluation, auditing, and other literatures) but rather was intended to be a targeted summary examination geared toward helping the Commonwealth of Australia redress contestability in defence settings.

literature on contestability per se, nor do departments and ministries of defence beyond the Commonwealth of Australia routinely use the term, which in other contexts is referred to as *scrutiny* or *challenge*. That said, a significant amount of work describes and analyses the concepts conveyed by the term. This literature enabled us to describe how contestability is used in different venues, both public and private, and to identify different elements that go into an effective contestability function.

Our review identified four governance functions—scrutiny, oversight, due diligence, and auditing—that organisations put in place to ensure and verify decisions about resource allocations. These four contestability functions form the basis for the contestability models and practices that public and private institutions typically follow. Each has specific circumstances and time frames, shown in Table S.1, for when the functions apply to governance.

These contestability functions offer a number of different formal ways to undertake independent reviews. Together with subject-matter expertise, we used more-detailed descriptions from the literature describing these functions to identify a dozen key aspects of contestability, which are displayed in Table S.2. We then used this list to inform and create the protocol that we employed in the second phase

Table S.1
The Four Contestability Functions

Name	Purpose	When It Occurs
Scrutiny	Function of a third party examining the process, information, and underlying assumptions used in an analysis or in the options presented for a major decision	Before decisions are made
Oversight	Less targeted than auditing; typically takes on a whole range of forms of authoritative supervision	Before, during, and after decisions
Due diligence	May be associated with compliance with legal and regulatory requirements	Before decisions are made
Auditing	Typically used in verifying past practice, including financial matters; used in both public and private sectors	After decisions have been made

Table S.2
Aspects of Contestability

Aspect	Accompanying Questions
Function	• What does the contestability function do? • What is the history of that function?
Institutionalisation	• Are the organisation and its contestability function defined in law, regulation, or policy?
Structure	• Where is the organisation located within the larger organisation that it serves? • Does placement in the hierarchy cause a conflict of interest? • How is the management of this office defined (e.g., appointed, elected, term)? • Does the contestability function have access to necessary data?
Type of engagement	• What is the duration or periodicity of contact between the contestability function and other offices? • Does this organisation have directed authority over capability developers, or does it have to go to a higher authority?
Funding	• What is the total funding? • How is this function funded?
Outputs and recipients	• What is the output of the contestability function (e.g., internal analysis, external analysis)? • Who is the final product for (e.g., senior leadership, the public)?
Standards	• How does this organisation develop and maintain contestability standards?
Staffing	• What is the staff composition and size of this function (e.g., experience, background, skill sets)? • How is the staff recruited? • How is the staff trained?
Incentives	• What incentives exist in this function to promote appropriate behaviour by the staff (e.g., financial incentives, promotions, publishing rights)? • What is appropriate staff behaviour in this office defined to be?
Organisational culture	• What are the shared values and norms that underpin behaviour?
Metrics	• What metrics are used to measure success of this function?
Risk	• How is risk handled? • What are the types of risk that are assessed (e.g., financial, technical)? • How does the organisation balance cost and capability trade-offs in a risk framework?

of our research, which involved discussions with public- and private-sector experts on contestability.

We found that the literature on contestability focuses primarily on oversight (by corporate boards) and on audit (corporate disclosures to shareholders and regulators, government financial disclosures to the public) functions. These approaches stem from the notion that making information public, or at least more widely available, reduces corruption and insular decisions.

We also found that the common thread through all aspects of contestability organisations is the avoidance of conflicts of interest. A contestability organisation has the least conflict of interest when it exhibits the following aspects:

- It cannot be eliminated arbitrarily (institutionalisation).
- It is not controlled by the people it reviews or analyses (structure).
- It cannot be left out of deliberations or decisions (type of engagement).
- It is not paid for by the people it reviews or analyses (funding).
- It is allowed to disseminate work products beyond the people being reviewed or analysed (outputs and recipients).

According to one report, efforts to minimise conflict of interest "serve two overarching purposes: maintaining the integrity of professional judgment and sustaining public confidence in that judgment."[5] Evidence in the financial sector suggests that recommendations are less biased[6] and more reliable from analysts who do not have a conflict.[7] In addition, even if the contestability organisations' findings or recom-

---

[5]   Bernard Lo and Marilyn J. Field, eds., *Conflict of Interest in Medical Research, Education and Practice*, Washington, D.C.: Institute of Medicine of the National Academies, 2009, p. 49.

[6]   Matthew L. A. Hayward and Warren Boeker, "Power and Conflicts of Interest in Professional Firms: Evidence from Investment Banking," *Administrative Science Quarterly*, Vol. 43, No. 1, March 1998.

[7]   Roni Michaely and Kent L. Womack, "Conflict of Interest and the Credibility of Underwriter Analyst Recommendations," *The Review of Financial Studies*, Vol. 12, No. 4, 1999.

mendations are not adopted, simply subjecting a decision to evaluation or review helps to increase the perceived legitimacy of the decision.[8]

Case Study Research and Discussions

In coordination with the sponsor, RAND selected a sample of public- and private-sector organisations from around the world that have shown some involvement with contestability; collected data on organisations from a variety of sources, including their official websites; and held discussions with experts working there. On the public side, we looked at eight departments or ministries of defence, which were identified with the help of this study's sponsor, and held discussions either with experts on the contestability function in their home countries or with defence attachés at embassies in Washington, D.C. In addition, we researched and held discussions with two nondefence federal agencies in the United States, one of which requested anonymity and the other of which (the U.S. Government Accountability Office [GAO]) we investigated through a literature review and through insights gleaned from an outside expert. A short description of a third nondefence governmental agency, the UK's National Audit Office (NAO), is also included. Our research and discussions in the private sector involved three commercial companies.

We used a purposive sampling approach, focusing on government organisations and industrial sectors in which our sponsor had indicated interest. These entities provide a fairly broad panorama of contestability approaches taken by organisations involved in large acquisitions. Table S.3 shows the array of entities covered in this research.

We found that just as there are multiple ways to perform contestability, it is not uncommon for organisations to carry out multiple contestability functions at various levels or to execute those functions inside a larger organisation (e.g., within the UK Ministry of Defence [UK MoD] or the U.S. Department of Defense). Moreover, there are

---

[8]   Anders Hanberger, "The Real Functions of Evaluations and Response Systems," *Evaluation*, Vol. 17, No. 4, 2011, p. 344.

**Table S.3
Organisations Included in the Case Studies**

Government Defence Ministries, Agencies, Defence Attaché	Nondefence Government Agencies	Commercial Enterprises
• Canadian Department of National Defence • Denmark Ministry of Defence • German Ministry of Defence • Netherlands Ministry of Defence • New Zealand Ministry of Defence • Sweden Defence Attaché • UK Ministry of Defence • U.S. Department of Defense	• GAO[a] • A large U.S. contestability organisation in a U.S. agency that requested anonymity • NAO	• An international shipbuilding and transportation firm • An international conglomerate firm • An international security and aerospace firm

[a] This discussion was with an expert on contestability who had also served on a congressionally mandated panel chaired by the comptroller general of GAO.

contestability organisations, such as GAO, that are independent of the organisations that they audit.

Our case studies did not reveal contestability best practices among the organisations that we researched, because it was not possible to link specific practices to outcomes. However, we did find a pervasive understanding that contestability could be linked to better outcomes and that reviewing decisions could help reduce or avoid, although not entirely eliminate, problems. We also assert that there are summary principles of contestability that will enhance the success of the new function over time.

Summary Insights

Our data collection revealed that different organisations take a wide variety of approaches to implementing and conducting contestability functions. Each organisation has developed unique approaches to putting in place checks and balances that govern its decisions connected with large capital expenditures. Table S.4 summarises how the organisations we researched manage contestability. It lists the defence organisations by total defence budget, including an entry for Australia to show where it would stand in this group. The UK National Audit Office[9] and the two U.S. government agencies are next, followed by the three commercial firms.

Why Develop a Contestability Function?

Interviewees advanced several motivations for creating a contestability function. In some cases, the function simply evolved organically as a mechanism to provide better accountability to the customer and increase the quality of the decisionmaking process. In others, interviewees were spurred by reviews that called the function into being or resulted from the desire to strengthen existing approaches. Poor acquisition outcomes (e.g., schedule delays, cost overruns, and not meeting technical specifications) also led organisations to develop the function.

However, contestability reviews were not the only mechanism mentioned as a means of improving the decisionmaking process. All European Union countries, by law, operate under a "competitive" procurement regime specifically designed to promote price competition. Sweden, for example, mentioned that it values competition as a way of reducing risk in acquisition. However, Sweden often focuses on meeting specific needs and can buy its systems in the marketplace, thus generating price competition. Organisations that try to push the technological state of the art are more likely to engage in high-risk development, which may be more likely to fail and may increase the need for a more formal contestability function.

We note that having a robust contestability function does not necessarily guarantee successful acquisition procedures. Other factors can intrude. The Cost Assessment and Program Evaluation (CAPE) office in the United States has an extensive, rigorous, and technically

---

[9]   This is included for contrast—we had insufficient information to create a full case study.

Table S.4
Contestability Function Summary Characteristics

Organisation	Country Defence Budgets (FY 2014 AUD, Millions)[a]	Main Contestability Function Examined	History	Number of Staff	Stand-Alone Function	Direct Report to Decision-maker	Decision-making Authority	Type of Engagement	Outputs/Recipients
U.S. (CAPE)	803,744	Scrutiny	Originated 1961, current structure dates from 2009	160	Yes	Yes	No	Constant, ongoing	Mostly internal; Secretary of Defense/ Department of the Secretary of Defense
UK (UK MoD scrutiny function)	79,704	Scrutiny	20 years old	60, with 15 for technical and operational analysis scrutiny	Yes	Yes	No	Constant, ongoing	Internal; scrutiny; report to Investment Approvals Committee
Germany (BAAINBw and Rü Board)	61,218	Scrutiny	~2014	Fewer than 10	No (chairman and ad hoc board)	Yes	No	Constant, ongoing	Annual report to Parliament; assessment to Chairman of Rü Board
Australia	33,487								

Table S.4—Continued

Organisation	Country Defence Budgets (FY 2014 AUD, Millions)[a]	Main Contestability Function Examined	History	Number of Staff	Stand-Alone Function	Direct Report to Decision-maker	Decision-making Authority	Type of Engagement	Outputs/Recipients
Canada (IRPDA)	22,998	Scrutiny, oversight	Established in June 2015 (still in development)	10	Yes (formal panel and stand-alone supporting office)	Yes	No	Constant, ongoing	Memorandum from the panel chair via the deputy minister of national defence to the minister
Netherlands (DMO)	13,292	Scrutiny, oversight	Evolved from a process started in 1980s	No dedicated staff; other duty as assigned	No (ad hoc panel with potential future challenge function)	Yes	No	End of each major phase	Final report to Parliament and the minister
Sweden	8,662	Oversight	N/A	No dedicated staff; other duty as assigned	No (review as part of strategic and long-term planning)	No	No	Constant, ongoing	Internal and external; minister for defence and Riksdag
Denmark (DALO)	5,874	Audit, oversight	Established in 2006 as DALO	No dedicated staff; other duty as assigned	No (review as part of capability-development process)	Yes	No	As needed for large investments	Internal; defence minister

Table S.4—Continued

Organisation	Country Defence Budgets (FY 2014 AUD, Millions)[a]	Main Contestability Function Examined	History	Number of Staff	Stand-Alone Function	Direct Report to Decision-maker	Decision-making Authority	Type of Engagement	Outputs/ Recipients
New Zealand (Gate Review Panel)	3,175	Scrutiny, oversight	~2010	No dedicated staff; other duty as assigned; external assistance	No (ad hoc review panel)	Yes	Yes	Panel convenes at gate reviews	Internal; prime minister/ secretary of defence

Nondefence Organisations (public agencies and commercial firms)

Organisation	FY 2014 Revenue or Sales, AUD, Millions	Main Contestability Function Examined	History	Number of Staff	Stand-Alone Function	Direct Report to Decision-maker	Decision-making Authority	Type of Engagement	Outputs/ Recipients
UK (NAO)	N/A	Audit	Established in 1983, in function long before	800	Yes	Yes	No	Constant, ongoing	Internal/ external; reports to Parliament
U.S. (large agency's ICG)	N/A	Scrutiny, oversight	2005 (in current form)	No information	Yes	Yes	No	Constant, ongoing	Internal; head of agency
U.S. (GAO)	N/A	Audit	Founded in 1921	3,000	Yes (stand-alone agency)	Yes	No	Constant, ongoing	Internal, but mostly external for public; Congress

Table S.4—Continued

Organisation	FY 2014 Revenue or Sales, AUD, Millions	Main Contestability Function Examined	History	Number of Staff	Stand-Alone Function	Direct Report to Decision-maker	Decision-making Authority	Type of Engagement	Outputs/Recipients
International shipbuilding and transportation firm	1,307 (sales)	Scrutiny	No information	5 people from various outside firms	No (ad hoc—red-team sessions are about 3 times per year)	Yes	No	About 3 times per year	Report to senior leadership
International security and aerospace firm	60,000+ (net sales)	Oversight, audit	No information	Depends on scale of the decision	No (ad hoc strategic panel and program excellence team)	Yes	No	As needed for investments	Report to executive vice president
International conglomerate firm	60,000+ (revenue)	Scrutiny, audit, due diligence	8–10 years old (existed organically back to 1904)	Strategy (6–8); risk (6–8); finance/accounting (40–50); internal audit (10–15)	Yes (multiple enterprise level); ad hoc (lower level)	Yes	No	Constant, ongoing	Investment review report to CEO, CFO, board of directors

a Based on average daily currency exchange rate between 1 December 2014 and 1 December 2015 of 1.3178 AUD to 1 USD, from Oanda.com.

SOURCE: Military-spending figures from Stockholm International Peace Research Institute, SIPRI Military Expenditure Database, undated.

NOTES: BAAINBw = Bundesamtes für Ausrüstung, Informationstechnik und Nutzung der Bundeswehrs (Federal Office of Equipment, Information Technology and In-Service Support); IRPDA = Independent Review Panel for Defence Acquisition; DMO = Defence Materiel Organisation; DALO = Defence Acquisition and Logistics Organization; NAO = National Audit Office; ICG = Independent Cost Group.

demanding contestability organisation, and the U.S. Department of Defense still has significant cost overruns on many programs. The ultimate decisionmakers also have the responsibility to consider political, economic, social, and other considerations that are outside the purview of a defence contestability function, which may well override the recommendations made by the contestability function.

Insights on Aspects of Contestability

Table S.5 depicts our broad findings on the various aspects of contestability.

Based on our research, there is no clear "right" answer or best organisational practice in contestability. It is difficult to assert that one or another approach is best when outcome metrics are problematic because the desired performance relies on many factors outside the control of the contestability function. However, we would note that one of the reasons why it is so difficult to link contestability to better outcomes is that it is specifically designed to help avoid bad outcomes, and the results of the road not taken cannot be knowable.

Local context and culture does and should drive the specific approach to contestability. That said, we note that a number of factors did not vary across the organisations we studied. These are important aspects of any contestability function and should be part of the initial design:

- Ensure that participants have a sense of independence.
- Ensure that participants can offer their inputs without fear of retribution.
- Ensure that that adequate resources are provided so that all decisions reaching whatever threshold is used to require review can be reviewed, so that there is no chance for a biased selection of what to analyse.

Table S.5
Findings on Aspects of Contestability

Aspect	Insight
Function	• Function varies by organisation • Function focuses on militarily significant investments • No organisation reviews every single investment decision • Industry tends toward a red-team approach
Institutionalisation	• Most organisations are formed from policy decisions • Scrutiny functions have predated the existence of formal scrutiny organisations • Functions not seen as at risk
Structure	• Structure varies substantially in size • Biggest discriminator is stand alone or internal to another process • Stand alone varies: some organisations undertake reviews and some only facilitate them • Many contestability functions are common at different levels • Jointness[a] facilitates (or forces) trade-offs
Types of engagement	• Type of engagement varies as a function of what organisation was created to do: challenge decisions or improve decisionmaking processes • Organisations do not review all decisions
Funding	• Funding is not a common concern • Commercial red teams save more than they cost • No one has enough staff to do all analyses desired
Outputs	• Outputs include recommendations to others • Approve or cancel program investments (in some cases) • Outputs often can be overturned
Standards	• High-quality standards • Most defence contestability functions do not generate independent estimates for their reviews; they start with program/project staff input • Collection and storage of data is a challenge • Retroactive reviews are difficult (how do you know whether it was a good decision?)
Staffing	• Personnel are highly experienced • Staff are mostly civilian, but, if military, very senior • Use of external expertise is common • Most organisations are often staffed with analysts or outsiders with expertise in specific areas
Incentives	• Contributions to organisation mission are commonly cited • Financial incentives are never cited (even for red-team participants)

Table S.5—Continued

Aspect	Insight
Culture	• Independence valued
Metrics	• There are no standard metrics • Typical focus is to improve quality and robustness of decisionmaking and ensure that public funds are well spent • Metrics provide better defence of decisions
Risk	• This aspect is frequently used because senior leaders need to understand risks • Both public and private organisations take account of risk in making decisions, with financial risk the focus of commercial organisations

[a] The degree of integration among different military services.

In addition to the above themes required for a contestability organisation, defence leaders likely will face a number of questions as they develop an approach appropriate to the local defence context:

- What is the primary function of the contestability regime? What level of technical, financial, and other scrutiny is desired? Are these desired functions realistic?
- Should there be a standing contestability organisation with the responsibility of performing the reviews, or should there be a review structure with a small footprint that can pull teams together on an as-needed basis? If a standing organisation, should it be stand alone or incorporated into an existing function or structure?
- Will the organisation have a permanent staff large enough to do all the work in-house or will it sometimes need to engage external staff? Will the staff consist of government officials or contractors, or will it depend on what expertise is required? Or will it have a small staff that facilitates reviews and designs each review separately? How can it be ensured that this staff has the resources necessary to accomplish the mission?
- To whom does the contestability organisation report and what are the outputs? What is the nature of the interrelationships among

the oversight and decision authorities in the ADoD and national-level oversight (e.g., Parliament) and audit functions? To what extent does the contestability organisation make final decisions (e.g., to cancel programs) as opposed to advising higher-level decision authorities?

- Will the decisions be contested at the decision gates? Or will the organisation work to improve ongoing decisionmaking throughout? What thresholds will trigger a review?
- Will the contestability organisation be given unfettered access to all data, information, and reports collected during the initial decisionmaking process, or will it have its own independent data sources—or both? And will the organisation under review be required to respond explicitly to the contestability organisation's findings? Will it review outcomes over the longer term to see the impact of decisions, as a source of lessons learned for the function?
- Will the organisation play a role in maintaining and managing data?

How defence leadership answers these questions will determine the nature, direction, and size of Australia's contestability component. The key goal should be the creation of a capability that is independent and objective: Senior leaders need to embrace the notion of having independent reviews of decisions that have not been previously reviewed, and, for the function to be most effective, organisations need to be willing to accept challenges to their decisions and adapt as required.

Organisational Change

Adding a contestability organisation could bring about major organisational change, which is always a challenge. Effective change does not occur merely with the addition of a new box on the organisation's wiring diagram. This raises the question of how Australia can ensure that its new contestability organisation will succeed. We recommend that the establishment of such a function be treated as a significant organisational change and then be managed according to best practices in change management, a topic on which there is a substantial literature with well-understood lessons.

Several key aspects of the change effort are already in place. Australia has plans to go ahead with a contestability function, and the need for change is laid out in *First Principles Review: Creating One Defence*.[10] Structuring the change (the "what") will not necessarily be a simple task—indeed, this report is specifically aimed at providing support for that process and shows that there are a number of reasonable alternative structures and subprocesses to select from—but once these decisions are made, then the focus needs to be on the "how." Senior-management support—by which we mean support from leaders at the top of the ADoD, not individuals internal to the new contestability function—is crucial to any change. Senior-management backing will be required in setting up the function within the organisation and process structure, including the new structure and the review requirements, and is necessary to ensure support from the rest of the organisation. Types of support required include getting additional backing and buy-in from all layers of management, adequate resourcing in terms of having the right number and mix of employees, support when those employees pull in expertise from across the ADoD, and adequate resources to hire external support if needed. Also, management support must be consistent, with the goal of imbuing contestability in the culture and having all relevant organisations either agree that it is a value added step, or at least acceding to the necessity of doing it. This means that once the framework for review and the decisionmaking threshold are determined, the organisation must support these on an ongoing basis. Allowing exceptions for programs to avoid review will fundamentally weaken the function and create the danger that stakeholders will focus on arguing for exceptions rather than ensuring that their analyses are solid.

---

[10] ADoD, 2015.

Summary Principles

We close by offering some summary principles for the incorporation of a successful contestability function, in the text box.

There are many choices in contestability approach and organisational design. Specific details of strong contestability functions do vary, but these principles will enhance the success of the contestability function over time and will help create a strong and effective system of checks and balances to ensure the best allocation of public resources in the defence of the nation.

> ## Summary Principles for Incorporating Contestability
>
> Senior-leadership and line-manager support for new contestability function
>
> Clear mission and understanding of what resource decisions the contestability function will review and where it is injected into the process
>
> Clear understanding of the outputs and goals of the contestability reviews
>
> Ongoing leadership support for contestability recommendations as quality inputs are to be taken very seriously (even if final decisions are different)
>
> Whomever the contestability function reports to has real decisionmaking power
>
> Adequate resources for the contestability function
>
> Adequate staff of senior, experienced experts for the contestability function
>
> Timely access to the right data
>
> Development of an independent review culture, without fear of retribution
>
> Storage of analysis and decisions over time to create a body of knowledge, which will help to increase the long-term success of the function

Acknowledgments

This work could not have been undertaken without the special commitment and support that the Australian Department of Defence (ADoD) provided RAND. For that support over the several weeks of intensive study and analysis, we are grateful. Many individuals in the ADoD, in the U.S. government, and in other public- and private-sector organisations shared insights that were critical to our qualitative analysis and to the interpretations and conclusions described in this report. Their names and contributions would fill several pages.

We particularly wish to recognise and thank three individuals without whose broad-based participation and support this analysis would not have been possible: Marc Ablong, First Assistant Secretary, White Paper; Todd Mansell, Chief, Joint and Operations Analysis Division; and Gabrielle Burrell, Contestability Team.

RAND colleagues John Birkler, Roger Lough, and Jennifer Moroney provided wise guidance throughout the drafting of this report. Christopher Mouton and Laura Baldwin provided thoughtful management reviews of the draft. Dan Jenkins and Igor Mikolic-Torreira provided perceptive peer reviews of an earlier draft, and Laura Werber provided many valuable insights. Their careful reads of our document and insightful comments occasioned many changes that improved the substance and clarity of the final product.

In addition, we would like to thank Charles Nemfakos for his insights on contestability. Another RAND colleague, Mark Arena, was particularly helpful to the research team with respect to defining contestability activities. Ian Cook and Sarah Soliman provided a variety of thoughtful insights and research support.

The authors owe RAND colleagues Jack Riley, Nancy Pollock, Michelle Platt, and Olivia Cao thanks for their thorough and patient assistance at every stage in the project.

CHAPTER ONE

Introduction

Next to the conduct of war, one of the most important functions of a nation's defence organisation is to invest its military with capabilities that deal with future threats. This function oftentimes requires that defence organisations make major expenditures, which attract both public and political scrutiny to ensure that the public's money is spent wisely. However, ensuring that such expenditures are indeed wise can be difficult for a host of reasons: The guiding strategic environment is often not universally agreed upon; the normal return-on-investment measures that are found in the commercial world do not apply; and, inasmuch as military capabilities often press the envelope of available technologies and are inherently risky, cost and schedule overruns are not uncommon.

Almost all large organisations have checks and balances as they make decisions about large capital expenditures. Such checks and balances are usually built into the organisation's bureaucracy and are adjusted from time to time to take into account changes in circumstances, resourcing environments, or organisational culture. At other times, the processes by which these checks and balances are made become ossified and need a major shake-up.

The recently completed report *First Principles Review, Creating One Defence* criticised the current processes through which the Australian Department of Defence (ADoD) develops capabilities.[1] This criticism led the report to recommend (and the government to accept)

---

[1]   ADoD, *First Principles Review, Creating One Defence*, Canberra, April 2015.

major changes in the ADoD's organisational structures and in its decisionmaking responsibilities and processes.[2] One key recommendation called for the establishment of an arm's length contestability function in the ADoD to provide "assurance to the Secretary that the capability needs and requirements are aligned with strategy and resources and can be delivered."[3]

This function currently is being designed as part of the implementation of the *First Principles Review*. It is not simple. Besides the obvious issues connected with organisational structure, questions surround how to define the contestability function within the capability development and maintenance line-management context, how to sustain it, how to manage risk, and so on.[4] To help answer these questions, the RAND Corporation has been engaged by the ADoD to identify practices, procedures, and elements that have been successfully implemented in foreign departments of defence and related agencies and in large commercial organisations in Australia and abroad.

Study Approach

This report documents the results of RAND's international scan, conducted between 14 September and 23 October 2015. It covers a first-order search of the relevant literature, including documented case studies, together with RAND-initiated discussions with U.S., European,

---

[2]   Recent changes in Australia include the implementation of an external oversight board, consisting of former government managers and industry experts (Julian Kerr, "Australia Announces Major DoD Shakeup," *Jane's Defence Weekly*, Vol. 52, No. 20, April 1, 2015). The oversight board adds to efforts made to improve the quality of the acquisitions workforce, as well as the appointment of a chief operating officer (COO) for acquisitions. ("Australia Pushing Ahead with Defense Procurement Reform," *Defense Daily International*, Vol. 11, No. 34, November 20, 2009; Julian Kerr, "Australian DoD Faces Sweeping Reforms," *Jane's Defence Weekly*, Vol. 48, No. 34, August 10, 2011).

[3]   ADoD, 2015, p. 38.

[4]   Other questions raised in that review included the need for two heads of acquisition decisions, the wisdom of bringing in potentially inexperienced civilian oversight to handle questions of air and naval warfare, and the end of the independence of the Defence Materiel Organisation (DMO).

and global organisations, to determine the existence and workings of any arm's length contestability function that organisations might have. We have drawn out some common themes, including resourcing and success measures, that may be useful pointers in subsequent deliberations within the ADoD as it implements the *First Principles Review*.

RAND contacted ten defence organisations that were identified by our sponsor and held discussions with officials in eight of them—Canada, Denmark, Germany, the Netherlands, New Zealand, Sweden, the United Kingdom, and the United States. The officials included senior officials working in the contestability functions and defence attachés located near the U.S.-based project team. All defence organisations agreed to be identified as such.

Additionally, RAND contacted two U.S. government agencies and held discussions with one. It asked not to be identified in the research, so we refer to it in general terms. For the other (U.S. Government Accountability Office [GAO]), we drew from a literature review and internal RAND discussions with those familiar with the office. Research on the United Kingdom's contestability function resulted in the collection of some information on the UK National Audit Office (NAO), which we include for reference, although research on that organisation was not conducted at the same depth as the other case studies.

We reached out to seven commercial companies and held discussions with four, of which three offered enough information to be included as summary case studies. They asked not to be identified by name. We contacted these companies because they operate in industrial sectors that were of specific interest to our sponsor.

Research Tasks and Approach

To conduct this analysis, RAND employed a qualitative research approach involving a literature review, discussions with experts on contestability, and the creation of models of existing contestability functions to accomplish the following three research tasks.

Task 1: Conduct an international scan to identify relevant contestability models. This task involved reviewing literature on contestability to (1) define *contestability* as it pertains to the ADoD's goals and (2) identify a list of potential public- and private-sector contestability functions relevant to defence. We used the literature and input from RAND experts in defence acquisition and other public- and private-sector experts to assemble a detailed list of characteristics of those functions.

As we were completing those elements, we consulted with the ADoD and specialists within RAND to identify public- and private-sector organisations to reach out to. To conduct the discussions, we used the protocol presented in Appendix A. We paid specific attention to the countries identified by the sponsor where enquiries with the Australian embassies and high commissions were already in process. We also identified several private-sector companies to identify further lessons learned and best practices.

Task 2: Assess the results of the international scan. In this task, we analysed the data collected in task 1. Our assumption was that each case would vary along the different contestability aspects that we had identified in task 1. We looked to see whether organisations followed standard models or they followed a menu approach, adopting and adapting various factors to fit their local contexts.

Task 3: Develop initial findings for the sponsor's consideration. In this task, RAND integrated the results of tasks 1 and 2. This task distilled the features of the international models surveyed, including the strengths and challenges that we identified in tasks 1 and 2.

Organisation of This Report

This report is organised into four chapters. Chapter Two covers different models of reviews that could fit under a larger contestability rubric and describes the different aspects of these. Chapter Three describes the results of the expert discussions. Chapter Four consolidates these findings and offers inputs on the key issues that must be decided as the Australians move forward with a formal contestability function.

Appendix A reproduces the discussion protocol that we used to solicit information from experts. Appendix B discusses examples of contestability functions from the literature. Appendix C contains a summary of financial thresholds for contestability reviews, with currency translations into Australian and U.S. dollars, for reference.

CHAPTER TWO

Literature Review and Analytic Framework

This chapter describes the literature on different ways of thinking about contestability and pulls from that literature different important aspects and activities to include when considering whether and how to develop a contestability organisation. This literature, combined with discussions with subject-matter experts, informed the protocol we used as a basis for our discussions in the international scan of contestability organisations.

What is *contestability*? At a very basic level, the root word, *contest*, can range in meaning from *fight* or *challenge* to a more collaborative *engage* or *respond*. Here we use a definition of *contestability* that means the capability of internal processes to enable stakeholders to challenge, engage, and respond to decisions about the allocation of resources, including major acquisition efforts and force structure. In other words, contestability is a governance function focused on ensuring capability needs and requirements that not only align with strategy and resources but also can be delivered effectively.

We note that the use of the term *contestability* to refer to this reviewing function is somewhat specific to Australia. There is no extensive literature on contestability per se, nor do other departments and ministries of defence routinely use the term. However, a significant amount of work describes and analyses the concepts conveyed by the term, which includes scrutiny, oversight, auditing, due diligence, and perhaps even performance evaluation (which often has an internal focus and thus is not covered in this research). Increasing the presence of these versions of contestability into decisionmaking processes has been a constant refrain in the private and public sectors for decades. Aspects of contestability organisations can be viewed through the lenses of the scope of the organisations' functions, institutionalisation, structures, typical engagement patterns, funding sources, and outputs and output recipients.

In this chapter, we describe how contestability is used in different venues, both public and private. Our literature review also covers the different aspects of what makes up an effective contestability function and describes the range of options for those aspects. Note that this is not meant to be an extensive and systematic review of the entire range of relevant work (which would cover management theory, evaluation, auditing, and other literatures) but rather a targeted summary with the goal of developing a manageable list of key considerations in the design of a new contestability organisation.

Although virtually any advice from the ADoD to government can be subject to contestability, usually it encompasses two main areas: (1) ensuring that the force structure and portfolio of capability investments deliver on the policy objectives and strategic needs set by the government and (2) providing independent challenges on the scope, schedule, budget, risks, and technical aspects of key defence projects—i.e., acquisition contestability.

Given the above potential uses of contestability, the ADoD is focusing more on examples of contestability that have scrutiny as their function. In this chapter, we also include the literature on oversight, audit, and due diligence because we wanted to ensure that we had identified, to the extent possible, the most-important considerations for creating a contestability organisation. The ADoD is not explicitly planning to add audit or due diligence functions, but the literature on those areas may still prove instructive.

Different Functions for Contesting Decisions

Formal governance research focuses on ensuring that a "diverse group of people is engaged in activities that contribute to the achievement of organisational goals."[1] Challenges include the fact that even the most well-meaning actors have bounded rationality[2] that not only limits their ability to see the whole picture but also means that they select some subset of relevant data to use for decisionmaking. Another issue is that the members of organisations may have goals and incentives that are at odds with those of the total organisation. For example, program managers have incentives to keep their programs alive to manage successful efforts, even if the programs no longer contribute to the goals of the organisation. This incentive may skew program managers' perception of cost and technical risk and lead them to judge these as lower than they actually might be. The potential for problems is greatest when there are conflicts of interest, an increased distance of the decisionmaker from oversight, more-complex decisions to be made, and decisions where errors could have serious consequences.[3]

In this analysis of contestability, we have identified four specific governance functions that are put in place to ensure a public sector entity's credibility. These four *contestability functions* form the basis for the broad understanding of contestability that allows a narrower focus on scrutiny throughout the rest of this report. Table 2.1 displays relevant elements for the four functions—scrutiny, oversight, due diligence, and auditing.

These contestability functions are described in further detail below.

Scrutiny

Scrutiny is the function of a third party examining the process, information, and underlying assumptions used in an analysis or the options presented for a major decision. Scrutiny takes place before decisions

---

[1]   Billy J. Hodge, William P. Anthony, and Lawrence M. Gales, *Organisational Theory: A Strategic Approach*, 5th ed., Upper Saddle River, N.J.: Prentice Hall, 1996, p. 241.

[2]   Herbert A. Simon, *Administrative Behavior: A Study of Decision-Making Processes in Administrative Organizations*, 4th ed., New York: The Free Press, 1997.

[3]   The Institute of Internal Auditors, *Supplemental Guidance: The Role of Auditing in Public Sector Governance*, 2nd ed., Altamonte Springs, Fla.: The Institute of Internal Auditors, January 2012, p. 13.

Table 2.1
The Four Contestability Functions

Name	Purpose	When It Occurs
Scrutiny	Function of a third party examining the process, information, and underlying assumptions used in an analysis or in the options presented for a major decision	Before decisions are made
Oversight	Less targeted than auditing; typically takes on a whole range of forms of authoritative supervision	Before, during, and after decisions
Due diligence	May be associated with compliance with legal and regulatory requirements	Before decisions are made
Auditing	Typically used in verifying past practice, including financial matters; used in both public and private sectors	After decisions have been made

are made. The scrutiny part of contestability, as used in this study, does not have a universally accepted definition but instead has different names in various contexts, including *independent assessment, expert review, devil's advocate, alternative analysis, white hat hackers,* and *red teams.* The role is ultimately similar, regardless of the specific term: attempting to identify shortcomings, uncertainty, or weaknesses in a system or analytic results. Scrutiny's contribution to contestability is to counter the effects of "confirmation bias"—by one author's definition, "an inclination to retain, or a disinclination to abandon, a currently favoured hypothesis."[4] In other words, people are inclined to stick to a set of favoured conclusions about a decision—even in the face of opposing evidence—and scrutiny is explicitly intended to counter that tendency. Incorporating scrutiny is accomplished by, among other approaches, "training people to think of alternative hypotheses early in the hypothesis-evaluation process,"[5] explicitly identifying the key

---

[4]   Joshua Klayman, "Varieties of Confirmation Bias," in J. Busemeyer, R. Hastie, and D. L. Medin, eds., *Decision Making from a Cognitive Perspective,* New York: Academic Press, 1995, p. 386.

[5]   Raymond S. Nickerson, "Confirmation Bias: A Ubiquitous Phenomenon in Many Guises," *Review of General Psychology,* Vol. 2, No. 2, 1998, p. 211.

assumptions supporting an approach or hypothesis,[6] and using people who "are not members of the organisation being assessed."[7] Scrutiny (or independent assessment) improves risk management—or at least the decisionmakers' understanding of risk—by highlighting underlying assumptions, shortcomings in analysis, the level of uncertainty, and the connection (or lack thereof) between an investment decision and an organisation's larger strategic plans.

As noted above, more emphasis has been placed on scrutiny in recent years, and the defence acquisition sector is no different. For example, greater scrutiny of acquisition in the U.S. Department of Defense (U.S. DoD) was deemed so important that a law created the permanent position of the Director of Cost Assessment and Program Evaluation (usually referred to simply as CAPE), which effectively made permanent the long-standing Office of Program Analysis and Evaluation (PA&E), with expanded roles. Among other tasks, CAPE is responsible for ensuring "that the cost estimation and cost analysis processes of the Department of Defense provide accurate information and realistic estimates" and is required to "conduct independent cost estimates and cost analyses for major defence acquisition programs."[8] Another role for this function is to assess the technical maturity of key components to a program, as risk related to technical immaturity is a major component of cost and schedule slip.[9] Some of CAPE's roles had been part of the U.S. DoD for many years, in PA&E, but CAPE's existence was not fixed in law, nor was it required to submit reports annually to the U.S. Congress.

Recent RAND work for the U.S. DoD's Performance and Root Cause Analysis (PARCA) Office has highlighted the importance of systematically identifying the key assumptions made very early in the

---

[6]   Mark V. Arena and Lauren A. Mayer, *Identifying Acquisition Framing Assumptions Through Structured Deliberation*, Santa Monica, Calif.: RAND Corporation, TL-153-OSD, 2014.

[7]   MITRE Corporation, *Systems Engineering Guide*, McLean, Va., 2014, p. 257.

[8]   United States Code, Title 10, Section 2334, Independent Cost Estimation and Cost Analysis, December 26, 2013.

[9]   See, for example, GAO, *Best Practices: Better Management of Technology Development Can Improve Weapon System Outcomes*, Washington, D.C., GAO/NSIAD-99-162, July 1999.

acquisition program life cycle.[10] We term these assumptions *framing assumptions*.[11] A framing assumption is any explicit or implicit assumption that is central in shaping cost, schedule, or performance expectations, which is defined by four criteria: critical, not easily mitigated, foundational, and program specific.[12] For example, a framing assumption may be the anticipated ability to increase a technology's level of maturity, but a framing assumption is not the expectation that a contractor will perform well. One potential activity of a contestability office would be to identify and track these assumptions over acquisition program life cycles. When assumptions prove to be false or invalid, this would signal a need to revise plans and budgets.

In the private sector and intelligence realms, the scrutiny function is often carried out by analysts who examine the same information and may or may not develop alternative, and sometimes intentionally contrary, findings to force the strengthening of the original analysis.

Oversight

Oversight takes a range of forms, generally defined as some type of authoritative supervision. *Oversight* is a commonly used term for governmental responsibilities. According to *The Encyclopedia of Political Science*, "Oversight plays a potentially important function in a chain of accountability linking the public to public policy decisions."[13] Matthew Amengual further defines *oversight* as "a broad term used to describe a variety of actions related to management and supervision in accountability relationships."[14]

---

[10] Arena and Mayer, 2014.

[11] By making framing assumptions more explicit early in the program life cycle and tracking them, the U.S. Office of the Secretary of Defense (OSD) and the services may be able to better manage major risks to and expectations of programs. Also, framing assumptions may change over the course of execution, or new ones could be added.

[12] Arena and Mayer, 2014.

[13] George Thomas Kurian, ed., *The Encyclopedia of Political Science*, Washington, D.C.: CQ Press, 2011, pp. 1163–1164.

[14] Matthew Amengual, "Oversight," in Mark Bevir, ed., *Encyclopedia of Governance*, Thousand Oaks, Calif.: SAGE Publications, 2007, pp. 655–656.

The U.S. Congress, for example, is considered to have oversight responsibility for the U.S. government by requesting reports, holding hearings, and crafting legislation that funds the operation of government agencies and directs them to take (or not take) certain actions.[15] Within government agencies, the senior leadership often has an oversight role, especially at the time of major decisions, such as resource allocation. Oversight achieves the contestability goal by creating a forum for considering risk management and return on investment.

While these are examples of oversight, they are not inclusive of all cases, because oversight is shaped and adapted to the specific needs, often influenced by historic practices. Oversight can be more effective when coupled with other types of contestability. For example, when there is a link between scrutiny teams (which often lack authority to stop major investments) and oversight teams (which generally have some kind of project authority), the work of the scrutineers can enable better oversight. This is true for all four contestability functions.

Assurance is another form of oversight. Many European governments employ an assurance or oversight function to maintain project performance, as well as accountability oversight.

Due Diligence

Due diligence reviews are carried out before decisions are made. While they may be associated with compliance with legal and regulatory requirements, they also encompass broader aims of ensuring that analyses incorporate all relevant data and that results and recommendations have not been biased toward a certain outcome. Robert Sprague and Sean Valentine provide a detailed description of due diligence and an example of how it is used in practice:

> Due diligence is a standard of vigilance, attentiveness, and care that is often exercised in various professional and societal settings. The effort is measured by the circumstances under which it is applied, with the expectation that it will be conducted with a

---

[15] Walter J. Oleszek, *Congressional Oversight: An Overview*, Washington, D.C.: Congressional Research Service, February 22, 2010.

level of reasonableness and prudence appropriate for the particular circumstances.[16]

Due diligence assists contestability by increasing the efficiency of operations and providing greater insights for the decisionmaking process. In the world of corporate mergers and acquisitions, for example, "effective due diligence should uncover issues which might de-rail negotiations or may lead to the failure" of the merger.[17] In other words, due diligence helps ensure that decision preparation has complied with requirements up front, which increases efficiency, and that the decision itself is more informed because it is based on evidence or data that are gathered as comprehensively as possible.

Auditing

Auditing is typically thought of as a detailed review or examination of financial matters to improve operations done by certified public accountants who assist in audits of private-sector entities.[18]

Auditing in the public sector is covered extensively in management theory and political science; audits are done to "provide information about resource management and accounting to higher authorities and the electorate to enable them to have confidence that public resources have been used properly and effectively."[19] Patrick von Maravic also describes public auditing as a means to achieving accountability in government. He offers additional insight into auditing in the public sector:

> Auditing is a form of oversight, or examination from some point external to the system or individual in question. Technically,

[16] Robert Sprague and Sean Valentine, "Due Diligence," in Robert W. Kolb, ed., *Encyclopedia of Business Ethics and Society*, Thousand Oaks, Calif.: SAGE Publications, 2008.

[17] Duncan Angwin, "Mergers and Acquisitions Across European Borders: National Perspectives on Preacquisition Due Diligence and the Use of Professional Advisers," *Journal of World Business*, Vol. 36, No. 1, 2001, p. 35.

[18] B. J. Reed and John W. Swain, *Public Finance Administration*, 2nd ed., Thousand Oaks, Calif.: SAGE Publications, 1997, pp. 301–316.

[19] Alexander Gash and John Wanna, "Audit," in Mark Bevir, ed., *Encyclopedia of Governance*, Thousand Oaks, Calif.: SAGE Publications, 2007.

auditing is a form of verification by an independent body, which compares actual transactions with standard practices.[20]

Summary Aspects of Contestability

Based on a combination of the literature review[21] and our subject-matter expertise, we identified a number of aspects of these contestability functions worth considering when creating or defining a contestability function. These aspects appear in Table 2.2.

Defining Aspects of Contestability Organisations

The sheer volume of available information on contestability functions, only a small part of which is reproduced here, suggests that contestability writ large is viewed as a useful practice that contributes to effective decisionmaking. The next step is to consider how to incorporate those functions into an organisation. The manner in which contestability roles are performed, and *who* performs them, are what appear to make the difference. Further examination of what is done and who does it reveals multiple key aspects of how to create organisations for these purposes—namely, the specific organisations' functions (defined above), institutionalisation of the organisations, management structures and reporting chains, typical engagement patterns, funding sources, outputs and output recipients, standards, staffing, incentives, organisational culture, metrics, and risk and risk management.

Institutionalisation

Institutionalisation is the extent to which an organisation is either a fixed, enduring set of functions or an entity that can be eliminated, curtailed, or significantly altered based on the whims of one or a few people. Institutionalisation ranges from the contestability func-

---

[20] Patrick von Maravic, "Auditing," in Bertrand Badie, Dirk Berg-Schlosser, and Leonardo Morlino, eds., *International Encyclopedia of Political Science*, Thousand Oaks, Calif.: SAGE Publications, 2011.

[21] Appendix B includes some examples from the practitioner literature.

Table 2.2
Aspects of Contestability

Aspect	Accompanying Questions
Function	• What does the contestability function do? • What is the history of that function?
Institutionalisation	• Are the organisation and its contestability function defined in law, regulation, or policy?
Structure	• Where is the organisation located within the larger organisation that it serves? • Does placement in the hierarchy cause a conflict of interest? • How is the management of this office defined (e.g., appointed, elected, term)? • Does the contestability function have access to necessary data?
Type of engagement	• What is the duration or periodicity of contact between the contestability function and other offices? • Does this organisation have directed authority over capability developers, or does it have to go to a higher authority?
Funding	• What is the total funding? • How is this function funded?
Outputs and recipients	• What is the output of the contestability function (e.g., internal analysis, external analysis)? • Who is the final product for (e.g., senior leadership, the public)?
Standards	• How does this organisation develop and maintain contestability standards?
Staffing	• What is the staff composition and size of this function (e.g., experience, background, skill sets)? • How is the staff recruited? • How is the staff trained?
Incentives	• What incentives exist in this function to promote appropriate behaviour by the staff (e.g., financial incentives, promotions, publishing rights)? • What is appropriate staff behaviour in this office defined to be?
Organisational culture	• What are the shared values and norms that underpin behaviour?
Metrics	• What metrics are used to measure success of this function?
Risk	• How is risk handled? • What are the types of risk that are assessed (e.g., financial, technical)? • How does the organisation balance cost and capability trade-offs in a risk framework?

tion being defined by a country's constitution, to established in law (e.g., GAO), to created by policy.

Structure

Management structures define how the contestability functions are assigned and who evaluates the performance of the organisation's personnel. Organisational structures can often be summarised in organisational charts, and therefore are probably the well-understood aspect of organisational design. Types of management structures include traditional line management, separate committees or boards, and outside reviewers or consultants.

All of these management approaches[22] can serve contestability, if the contestability organisation has management support.[23] However, for true independence, it is also necessary that individuals undertaking the contesting function do not report to the originators of the analytic output being reviewed.

Types of Engagement

Engagement patterns describe how often the contestability organisation discusses work or presents findings. The engagements can be periodic based on time (e.g., quarterly), event driven (e.g., in advance of decision points or budget deliberations), or as needed. Any pattern of engagement can enable the organisation to contribute to contestability as long as the engagements are required within the decision and deliberation processes (i.e., the organisation cannot be excluded when its input may run counter to the desired outcomes of others).[24]

---

[22] Thomas Asare, "Internal Auditing in the Public Sector: Promoting Good Governance and Performance Improvement," *International Journal on Governmental Financial Management*, Vol. 9, No. 1, 2009, p. 18.

[23] Anders Hanberger, "The Real Functions of Evaluations and Response Systems," *Evaluation*, Vol. 17, No. 4, 2011, p. 332.

[24] Robert Smith, *Audit Committees Combined Code Guidance: A Report and Proposed Guidance by an FRC-Appointed Group Chaired by Sir Robert Smith*, London: Financial Reporting Council, January 2003, p. 10.

Funding

The funding source is the primary authority controlling an organisation's budget. The legislative body is the ultimate source of funding in most governments, but in this context, the funding source is the authority or management component controlling the funding after the first step of appropriating money.[25] The funding source—and stability of that source—is important to contestability because explicit and implicit incentives tied to an organisation's budget can bias analyses. Concerns have been raised that audit conflicts may be resolved in the favour of large clients in the private sector, who typically pay larger fees.[26]

Outputs

Outputs of the contestability functions and those who receive the outputs indicate the extent to which evidence, analysis, and findings are disseminated outside the organisations, programs, and activities under review. The outputs can range from consultations with senior leadership to submissions of formal reports, with recipients ranging from internal management, to other government bodies (e.g., the legislature), to the public. Outputs and recipients contribute to contestability by increasing the audience and providing opportunities for further engagement on major decisions.[27]

Note that there is a distinction between *outputs*, which are indicators of activities, and *outcomes*, which are measures of results. We were unable to link the outputs of contestability models to outcomes in the defence organisations, where desired outcomes tend to be very macro—for example, "keeping the nation safe." These outcomes are subject to many factors well beyond the contestability function's control. However, we would note that one reason why it is so difficult to

---

[25] Isabelle Bourgeois, Eleanor Toews, Jane Whynot, and Mary Kay Lamarche, "Measuring Organisational Evaluation Capacity in the Canadian Federal Government," *The Canadian Journal of Program Evaluation*, Vol. 28, No. 2, 2013, p. 10.

[26] Donald R. Deis, Jr., and Gary A. Giroux, "Determinants of Audit Quality in the Public Sector," *The Accounting Review*, Vol. 67, No. 3, July 1992, p. 466.

[27] Smith, 2003, p. 10.

link contestability to better outcomes is that it is specifically designed to help avoid bad outcomes, and the results of the road not taken cannot be knowable.

Standards

Another, perhaps less obvious, function of a contestability office is to develop standards, methods, and tools for key program plans, such as cost and schedule estimates.

The U.S. DoD's CAPE has requirements for the structure (work breakdown structure) for all life-cycle cost estimates.[28] Some organisations have developed tools to standardise work and process, assess program maturity at various milestones,[29] and evaluate systems engineering risk.[30]

Larger commercial firms also develop similar tools and guidelines. Moreover, they may have explicit guidelines in terms of the program-development and -execution processes.

The standards for an oversight function are also based on the organisation's mission. For instance, GAO conducts oversight on behalf of the U.S. Congress. Its mission is to

> support the Congress in meeting its constitutional responsibilities and to help improve the performance and ensure the accountability of the federal government for the benefit of the American people. We provide Congress with timely information that is objective, fact-based, nonpartisan, non-ideological, fair, and balanced. . . . Our Core Values of accountability, integrity, and reliability are reflected in all of the work we do. We operate under strict professional standards of review and referencing; all facts and analyses in our work are thoroughly checked for accuracy. In addition, our audit policies are consistent with the Fundamental

---

[28] CAPE, *Operating and Support Cost-Estimating Guide*, Washington, D.C., March 2014.

[29] Naval Air Warfare Center, Training System Division, "Technical Reviews," web page, April 8, 2015.

[30] Lauren A. Mayer, Mark V. Arena, and Michael McMahon, *An Excel Tool to Assess Acquisition Program Risk*, Santa Monica, Calif.: RAND Corporation, TL-113-OSD, 2013.

Auditing Principles (Level 3) of the International Standards of Supreme Audit Institutions.[31]

Oversight standards depend on the nature of the oversight. At the same time, "auditing standards are professional guidelines promulgated either by an authorized national or international body."[32]

However, given the still-developing nature of scrutiny, the literature does not contain research specific to the scrutiny function, and, consequently, broadly accepted standards do not exist. Nevertheless, the standards are important to think about and would likely be defined by the organisation requesting the analysis and those potentially conducting the analysis.

Staffing

The staff of a contestability function is one of the key parts of that function's success. The right composition, background, skill set, size, and training are some of the factors to consider for staffing a contestability function.

In the literature, there are no recommendations on staffing size for scrutiny functions, in part because of the difference in the functions, the structure of the functions, and the requirements for review. Large organisations with a formal function and many decisions to review may benefit from a permanent staff. Others may staff the function on a rotating basis, given the nature of the decision, or pull in red teams or "tiger teams" from outside to conduct the reviews. Some organisations may pull in contractors or other subject-matter experts on an as-needed basis.

Incentives

In the management literature, discussions of incentives largely link to monetary rewards or rewards with monetary value, "including a premium wage system, but also nonfinancial rewards, such as the reduc-

---

[31] GAO, "About GAO," web page, undated-a.

[32] Sigmund Wagner-Tsukmoto, "Scientific Management," in Eric H. Kessler, ed., *Encyclopedia of Management Theory*, Thousand Oaks, Calif.: SAGE Publications, 2013, p. 679.

tion of work time, the provision of educational and recreational facilities, housing facilities, and other benefits."[33] However, "rewards can take many different forms, including praise from superiors and co-workers, implicit promises of future promotion opportunities, feelings of self-esteem that come from superior achievement and recognition, and current and future cash rewards related to performance."[34]

Monetary incentives may be harder to provide in the public sector because of formal pay scales, and contestability organisations are especially challenging because the desired outcomes are not necessarily a simple increase in productivity—i.e., "doing more" contestability work. In general, "performance in most jobs cannot be measured objectively because joint production and unobservability mean that individual output is not readily quantifiable,"[35] thus complicating how to design and structure incentives.

The research indicates that these considerations are further complicated by "the importance of weighing the influence of organisational structure and complexity, policy choices and constraints, and service-delivery practices in assessing program performance."[36] Ultimately, incentives for the contestability workforce must be carefully tailored to organisational and individual goals within the constraints of the government agency's broader mission.

Organisational Culture

Organisational culture is "a system of shared values (that define what is important) and norms that define appropriate attitudes and behaviors for organizational members (how to feel and behave)."[37] For any

---

[33] Wagner-Tsukmoto, 2013, p. 679.

[34] George P. Baker, Michael C. Jensen, and Kevin J. Murphy, "Compensation and Incentives: Practice vs. Theory," *The Journal of Finance*, Vol. 43, No. 3, July 1988, p. 594.

[35] Baker, Jensen, and Murphy, 1988, p. 597.

[36] Carolyn J. Heinrich, "Outcomes-Based Performance Management in the Public Sector: Implications for Government Accountability and Effectiveness," *Public Administration Review*, Vol. 62, No. 6, November/December 2002, p. 721.

[37] C. A. O'Reilly III, and J. A. Chatman, "Culture as Social Control: Corporations, Cults, and Commitment," *Research in Organizational Behavior*, Vol. 18, 1996.

organisational function, organisational culture shapes and is shaped by how the workforce within the contestability function perceives its mission, duties, and responsibilities. The culture can also be affected by how the larger organisation perceives the contestability function. An organisation conducting a contestability function cannot be perceived as an advocate because both independence and a supporting culture are critical for effective contestability. The greater organisation must take the independence of the contestability function seriously. Building credibility, a solid reputation of unbiased analysis, and the perception that the contestability function preserves the greater health of the organisation as a whole is important to the contestability function's success; this is because the culture of the overall organisation will understand and react appropriately to the contestability function's role in that organisation rather than trying to undermine the goals of the contestability function.

The World Bank provides some best practices for internal government audits, in which the World Bank focuses on *management support*. Though not specifically a scrutiny function, the qualities of management support in an internal audit function are similar:

> It is generally accepted that, to be effective, the internal audit function must have the full support of the organisation's senior management. The support of line management is also critical. The attitude of management towards internal audit can have a significant influence on the behaviour of an organisation's staff— similarly the attitude of management towards internal audit can either strengthen or hamper its role. . . . Internal audit needs to be pro-active in this respect both to set an example and to indicate better practice. This approach will both enhance its credibility and provide greater assurance to its stakeholders.[38]

In the private sector, corporate compliance has benefitted from the incorporation of compliance officers, who provide a type of contestability function. Michael Greenberg notes the following regarding

---

[38] World Bank, *Best Practices for Internal Audit in Government Departments*, undated.

the types of cultural aspects that need to be in place for that function to succeed:

> An empowered and independent CECO [chief ethics and com-pliance officer] is a basic element in any effective compliance program, and a key resource for boards in carrying out their oversight responsibilities. Strong compliance programs tend to involve a mix of both hard and soft elements—modifying the structure and control processes within firms, as well as seeking to promote culture changes. . . . [C]ompliance tends to work best when . . . there is no inconsistency between values and behavior, between internal and external messaging, and between the tone at the top and the controls and everyday practices throughout an organisation.[39]

Support by the contestability function's leadership and the over-all organisation's leadership is also important for the success of a con-testability function in both the public and private sectors. The senior decisionmakers have the final decisionmaking authority, so if they give credence to the contestability function, then it will have the power to improve decisionmaking:

> From an organisational standpoint one major element that sets apart successful from unsuccessful organisations is leadership, which should be dynamic and effective. Leadership can be seen

---

[39] Michael D. Greenberg, *Transforming Compliance: Emerging Paradigms for Boards, Man-agement, Compliance Officers, and Government*, Santa Monica, Calif.: RAND Corporation, CF-322-CCEG, 2014, pp. 1–2. Greenberg also says:

> Over the past six years, the RAND Corporation has organized a series of roundtable symposia, with the aim of exploring the intersection of compliance and governance, the exogenous factors that help to shape them, and related policy. The symposia have brought together accomplished thought leaders across a range of professional backgrounds and perspectives to grapple with new challenges and opportunities facing CECOs, boards, and senior management. Some of the topics covered have included the role of corpo-rate directors in compliance oversight, the implications of internal and external whistle-blowers for compliance risk, the multifaceted relationship between organizational cul-ture and compliance, and the unique difficulties associated with C-suite–level ethical lapses and compliance problems. Across these varied topic areas, several common themes have been raised by participants.

as the activity to influence others to willingly achieve specified objectives; it is clearly dependent on individual behaviours and a set of attributes which characterize a leader.[40]

Levi Nieminen and Daniel Denison support some of the above themes on organisational culture, and they provide ways to promote cultural effectiveness that could assist in the success of a contestability function:

> [T]he highest performing organisations find ways to empower and engage their people (involvement), facilitate coordinated actions and promote consistency of behaviors with core business values (consistency), translate the demands of the organisational environment into action (adaptability), and provide a clear sense of purpose and direction (mission).[41]

The U.S Office of Personnel Management provides a set of qualifications for effective leaders at the Senior Executive Service level, which is the level at which a person would lead a division within the contestability function in the U.S. DoD (e.g., CAPE, which is led by a political appointee). These executive core qualifications (ECQs) are as follows:

- *Leading change* involves the ability to bring about strategic change, both within and outside the organisation, to meet organisational goals. Inherent to this ECQ is the ability to establish an organisational vision and to implement it in a continuously changing environment.
- *Leading people* involves the ability to lead people toward meeting the organisation's vision, mission, and goals. Inherent to this ECQ is the ability to provide an inclusive workplace that fosters the development of others, facilitates cooperation and teamwork, and supports constructive resolution of conflicts.

---

[40] Levi Nieminen and Daniel Denison, "Organisational Culture and Effectiveness," in Eric H. Kessler, ed., *Encyclopedia of Management Theory*, Thousand Oaks, Calif.: SAGE Publications, 2013, pp. 530–531.

[41] Nieminen and Denison, 2013.

- *Results driven* involves the ability to meet organisational goals and customer expectations. Inherent to this ECQ is the ability to make decisions that produce high-quality results by applying technical knowledge, analysing problems, and calculating risks.
- *Business acumen* involves the ability to manage human, financial, and information resources strategically.
- *Building coalitions* involves the ability to build coalitions internally and with other federal agencies, state and local governments, nonprofit and private-sector organisations, foreign governments, or international organisations to achieve common goals.

ECQ are enhanced by a strong background in the following six competencies: interpersonal skills, public-service motivation, oral communication, integrity and honesty, continual learning, and written communication.[42]

Generally, leadership and culture focused on contestability do not appear spontaneously. Leaders of a contestability function must drive the development of a culture that understands how contestability benefits the larger organisation overall, and the top echelons of the larger organisation must explicitly support the contestability leadership, especially when contestability is being newly incorporated.

Metrics

A significant challenge in all public-sector organisations is how to develop appropriate metrics to assess organisational effectiveness. Developing these is especially difficult for contestability because "success" can take many forms, and managers face the problem of proving the negative—i.e., demonstrating how a decision or outcome would have been worse without the contestability work. Carolyn Heinrich summarised public-sector work:

> Research on performance management suggests that, in responding to the requirements of Government Performance and Results

---

[42] Cheryl Ndunguru, *Executive Core Qualifications: Becoming an Effective Leader, Senior Executive Resources and Performance Management, Training and Executive Development*, Washington, D.C.: U.S. Office of Personnel Management, undated.

Act, federal agencies should choose performance measures that (1) are closely aligned with their stated goals; (2) approximate actual performance as closely as possible; (3) are relatively simple and inexpensive to administer; and (4) make it difficult for managers to increase their measured performance in ways other than increasing their actual performance. [However,] [w]hen multiple or conflicting goals motivate employees, when organisational goals and performance measures diverge, or when bureaucratic effort across government levels is not readily observed, problems in performance-management systems are likely to arise.[43]

GAO's website on performance measures, for example, notes: "Many of the benefits resulting from our work cannot be measured in financial terms. These benefits can result in better services to the public, changes to statutes or regulations, or improved government operations."[44]

Benefits that are not easily quantified also may not be seen for many years. Deloitte's Advanced Analytics and Modelling practice reviewed GAO recommendations[45] over the course of 30 years in an effort to quantitatively determine whether GAO recommendations were an effective way to drive targeted change within agencies. Of

---

[43] Heinrich, 2002, p. 714.

[44] GAO, "Performance Measures," web page, undated-d.

[45] According to Deloitte,

> GAO has been able to quantify agency compliance by diligently following up year after year with agencies to see if recommendations have come to fruition. GAO's consistent and persistent approach underpinned our ability to conduct the multi-decade analysis above. Agencies which hope to quantify the success of their own internal oversight initiatives will need to commit to a similar level of effort.

> Tracking the success of recommendations is not only crucial for gaining an overall view of the oversight's efficacy but also improves the insights text analytics can produce. For example, if an agency clearly defines success for their recommendations, they would not only know how the contents of their recommendations had changed over time but would understand how the contents of their successful recommendations had changed.

Daniel Byler, Steve Berman, Vishwa Kolla, and William D. Eggers, *Accountability Quantified: What 26 Years of GAO Reports Can Teach Us About Government Management*, Westlake, Tex.: Deloitte University Press, 2015, p. 5.

the "40,000-plus recommendations made by GAO to federal agencies from 1983 through 2014," Deloitte found that "GAO recommendations have an 81 percent success rate."[46]

In examining the lack of research that addresses the quality of audits in contestability efforts, Jere Francis concludes that part of the problem is that

> Over time, the norms of science have increasingly emphasized [rigor of research] to the exclusion of [the intrinsic importance of research]. As a consequence, we tend to research those topics we can research rigorously. As a result, a lot of intrinsically important research questions do not get asked in the first place.[47]

Other research has shown how the contestability functions need to be adjusted to the tasks and environment, which affect the metrics. In studying management oversight of exploratory work, Rita Gunther McGrath asserted: "[M]anagement oversight that is pervasively success seeking in a narrow sense can lead an organisation to generate insufficient variety. This happens because deviation from plan in a negative direction can be branded a failure, which tends to squeeze out variety and inhibit learning."[48]

Thus, government agencies, generally—and contestability organisations, specifically—face the combined problem of multiple factors outside individual control influencing performance, benefits of government work accruing in difficult-to-measure forms (e.g., "better services to the public"), and a lack of research on how to improve performance. Also, contestability work is often viewed as an interruption or intrusion by the organisation being contested, which further complicates the notion of how to measure successful contestability.

Benchmarking is the process of comparing the performance and practices of other organisations, and a contestability organisation could

---

[46] Byler et al., 2015, p. 3.

[47] Jere R. Francis, "A Framework for Understanding and Researching Audit Quality," *AUDITING: A Journal of Practice and Theory*, Vol. 30, No. 2, May 2011, p. 145.

[48] Rita Gunther McGrath, "Exploratory Learning, Innovative Capacity and Managerial Oversight," *The Academy of Management Journal*, Vol. 44, No. 1, February 2001, p. 128.

assess whether a program's cost and schedule expectations are competitive or represent good value. This comparative process is used frequently in the commercial sector to identify strengths, weaknesses, and areas for improvement, although organisational dissimilarities can create a challenge.

The sine qua non of any contestability function is data, and a new contestability organisation will have to put in significant effort to collect data for the assessments, as well as to potentially normalise historical data. Many organisations encounter the problem of data being unstructured (e.g., it is only available in paper, PDF, and PowerPoint), presented in various formats, and collected inconsistently over time. Unstructured and inconsistent data severely limit the ability of analysts to provide the best insights. For example, RAND has spent considerable effort to collect and normalise Selected Acquisition Report data across major acquisitions, going back several decades.[49] This work involved manually extracting information from paper (or PDF) records to create a structured, aggregated database that could be analysed.

An example of the importance of data for contestability can be found in a recent GAO report on improving the portfolio management of the U.S. DoD's weapon system acquisitions.[50] GAO found:

> The Department of Defense (DOD) is not effectively using portfolio management to optimize its weapon system investments, as evidenced by affordability challenges in areas such as shipbuilding and potential duplication among some of its programs. Best practices recommend assessing investments collectively from an enterprise-wide perspective and integrating requirements, acquisition, and budget information.

The report continues:

---

[49] This data set has been continuously updated, but an initial description was published in Jeanne M. Jarvaise, Jeffrey A. Drezner, and Daniel M. Norton, *The Defense System Cost Performance Database: Cost Growth Analysis Using Selected Acquisition Reports*, Santa Monica, Calif.: RAND Corporation, MR-625-OSD, 1996.

[50] GAO, *Weapon System Acquisitions: Opportunities Exist to Improve the Department of Defense's Portfolio Management*, Washington, D.C., GAO-15-466, August 2015.

Both AT&L [Acquisition, Technology and Logistics] and the Joint Staff said that a lack of readily accessible data and analytical tools also hampered their ability to effectively conduct portfolio reviews. . . . Joint Staff officials said that past efforts relied on repeated data calls, which were a drain on resources and time consuming. In addition, one of the officials in charge of a Joint Staff portfolio review said that the Joint Staff does not have the authority to compel other DOD components to provide the necessary information and data to conduct the reviews. Finally, Joint Staff officials report that they do not have the analytical tools and dynamic databases to effectively conduct portfolio reviews, assess potential redundancy, and prioritize capabilities.[51]

In the commercial world, consulting firms gather program data in specific industries.[52] The objective is to identify both good and bad practices that affect performance and to consider guidelines based on those observations.

Risk and Risk Management

Organisations take into account risk when planning to use resources. They also monitor risk as projects or programs are being implemented. According to Michel Leseure,

Risk is the set of events and consequences, foreseeable or not, that occur, within or beyond the boundaries of an operations system, in reaction to the implementation of a plan of action. . . . Simply put, modern risk management is concerned with organisational exposure to volatility. For example, a manufacturer may suffer from fluctuating raw materials prices, or a banker may suffer from fluctuating default rates on loans.[53]

---

[51] GAO, 2015, summary and p. 23.

[52] See the websites for Independent Project Analysis (www.ipaglobal.com) or Asset Performance Networks (www.ap-networks.com) for examples in the oil and chemicals industries.

[53] Michel Leseure, "Risk," in *Key Concepts in Operations Management, Sage Key Concepts,* Thousand Oaks, Calif.: SAGE Publications, 2010.

In the world of advanced research in military investments, risk is always a consideration because of inherent technical limits. An important benefit of contestability functions is to help understand risk in a more objective manner that is not clouded by other considerations, such as vested stakeholder interests. Risk also needs to be considered in the context of trade-offs. In other words, while an investment may have a level of risk acceptable to management, by itself, the investment may be too risky when compared with competing demands and budgetary constraints.

Summary and Ways to Incorporate Contestability

There is a range of functions (audit, oversight, due diligence, and scrutiny) and important aspects of contestability. The literature on contestability in the private and public sectors focuses primarily on oversight (by corporate boards) and audit (corporate disclosures to shareholders and regulators, as well as government financial disclosures to the public) functions. These approaches stem from the notion that making information public, or at least more widely available, reduces corruption and insular decisions.

The common thread through all aspects of contestability organisations is conflict of interest. A contestability organisation has the least conflict of interest when it

- cannot be eliminated arbitrarily (institutionalisation)
- is not controlled by the people it reviews or analyses (structure)
- cannot be left out of deliberations or decisions (type of engagement)
- is not paid by the people it reviews or analyses (funding)
- is allowed to disseminate work products beyond the people responsible for the decision being reviewed or analysed (outputs).

According to one author, efforts to minimise conflict of interest "serve two overarching purposes: maintaining the integrity of professional judgment and sustaining public confidence in that judgment."[54] In other words, organisations characterised by these low conflict-of-interest aspects are viewed as more "independent"[55] or "objective" and are ascribed a greater degree of credibility in the contestability domain. Evidence in the financial sector suggests that recommendations are less biased[56] and more reliable from analysts who do not have a conflict.[57] In addition, even if the contestability organisations' findings or recommendations are not adopted, simply subjecting a decision to evaluation or review helps increase the perceived legitimacy of the decision.[58]

The literature review in this chapter was used to inform a discussion protocol (see Appendix A), as well as identify some activities to consider when developing and scoping a contestability function. Chapter Three is the output of discussions on how specific organisations conduct contestability in practice—with a primary focus on scrutiny—supplemented with publicly available information on these organisations for context.

---

[54] Bernard Lo and Marilyn J. Field, eds., *Conflict of Interest in Medical Research, Education and Practice*, Washington, D.C.: Institute of Medicine of the National Academies, 2009, p. 49.

[55] Asare, 2009, p. 17.

[56] Matthew L. A. Hayward and Warren Boeker, "Power and Conflicts of Interest in Professional Firms: Evidence from Investment Banking," *Administrative Science Quarterly*, Vol. 43, No. 1, March 1998.

[57] Roni Michaely and Kent L. Womack, "Conflict of Interest and the Credibility of Underwriter Analyst Recommendations," *The Review of Financial Studies*, Vol. 12, No. 4, 1999.

[58] Hanberger, 2011, p. 344.

CHAPTER THREE
Case Studies

This chapter provides brief summaries of 14 contestability case studies. As described previously, we used a purposive sampling approach, focusing on government organisations and the industrial sectors in which our sponsor had indicated interest. On the public side, we conducted open-source research on eight departments or ministries of defence (MoDs) and held targeted discussions either with experts on the contestability function in their home countries or with defence attachés at embassies in Washington, D.C. We researched and held discussions with two nondefence federal agencies in the United States—one requested anonymity and the other (the U.S. GAO) we investigated through a literature review and through insights gleaned from an outside expert. A short description of a third nondefence governmental agency, the UK NAO, is also included. Our research and discussions in the private sector involved three commercial companies.

The open-source research was conducted by drawing on publicly available articles, various governmental websites and publications, and historical financial and procurement or acquisition data. The targeted discussions the research team held with experts were guided by the protocol provided in Appendix A; however, each discussion proceeded in an organic fashion, based on the specifics of the organisation and the insights of our interviewee. Each case study differs in structure because some topics were not relevant to organisations without a formal structure for managing contestability. These case studies are intended to be short illustrative examples focused on contestability, rather than "thick" in-depth descriptions of the entire decisionmaking process.[1]

We translated all budget numbers into Australian dollars to facilitate easier comparisons across the summaries, using the average exchange rates for the year between 1 December 2014 and 1 December 2015. All translations are footnoted with the exchange rate used, and a summary table of major conversions is presented in Appendix C.

Public Organisations

Government Ministries and Departments of Defence

Government ministries and departments of defence summaries are presented in order of the size of the defence budget, from greatest to least.

United States

The U.S. DoD CAPE office provides an independent analysis and contestability function for the U.S. military. CAPE sits outside the functions that make procurement and force structure recommendations and reports to the Deputy Secretary of Defense.

CAPE sees its culture as its greatest strength and source of independence. All employees understand the importance of their independence. Although the head of the office is a political appointee, the office is strongly apolitical. Its role is that of the independent arbiter of options for the deputy secretary. Chartered under the 2009 Weapon Systems Acquisition Reform Act, but originating decades before in the Office of Program Analysis and Evaluation, CAPE and its predecessor agencies served the secretary, providing answers directly to the top.

RAND talked to a current official from CAPE and drew on other knowledge of the organisation.[2] The discussion highlighted CAPE's strong, independent analysis role within the U.S. DoD and offered the insights about CAPE's functions, workforce, outputs, and outcomes.

---

[1] GAO offers a useful framework for different types of case studies in *Case Study Evaluations*, Washington, D.C., GAO/PEMD-91-10.1.9, November 1990.

[2] Disclosure: CAPE is a funder of RAND research.

Functions

CAPE has three major scrutiny roles. The first is to produce independent cost estimates of major acquisition programs. By law, CAPE's analysis is not simply a verification of another office's analysis or a look at the varied cost estimates of a given program, but rather an independently derived estimate of what the real cost will be and whether it fulfils the requirements set by strategy. According to U.S. DoD Instruction 5000.02: "Director of Cost Assessment and Program Evaluation (DCAPE) provides policies and procedures for the conduct of cost estimates and cost analyses for all DoD acquisition programs, including issuance of guidance relating to program life-cycle cost estimation and risk analysis; reviews cost estimates and cost analyses conducted in connection with major defence acquisition programs (MDAPs) and major automated information system (MAIS) programs; and leads the development of DoD cost community training."[3] Specific cost estimates are conducted on certain program types and at certain times in the acquisition process:

- Cost estimates are conducted for MDAPs and MAIS programs for which the Under Secretary of Defense for Acquisition, Technology and Logistics (USD[AT&L]) is the Milestone Decision Authority (MDA), as well as requested by the MDA for other MDAPs and MAIS programs.
- Cost estimates are conducted for Acquisition Category (ACAT) IC and IAC programs at any time considered appropriate by the DCAPE or upon the request of the USD(AT&L) or the MDA.[4]
- For the MDAPs for which the DCAPE does not develop an Independent Cost Estimate (ICE), the ICE supporting a milestone review decision will be provided to the MDA by the applicable service cost agency or defence agency equivalent following review and concurrence by the DCAPE.

---

[3] U.S. Department of Defense Instruction 5000.02, *Operation of the Defense Acquisition System*, Washington, D.C.: Under Secretary of Defense for Acquisition, Technology, and Logistics, January 7, 2015, p. 127.

[4] An ACAT IAC acquisition program is a Major Automated Information System for which the Milestone Decision Authority is the DoD component head or, if delegated, the DoD component acquisition executive.

- DCAPE representatives will meet with representatives from the service cost agency and program office no later than 180 calendar days before the scheduled Development Request for Proposals Release Decision Point to determine what cost analysis, if any, will be presented at the decision review and who will be responsible for preparing the cost analysis.
- DCAPE reviews all estimates for MDAPs and MAIS programs, including estimates of operating and support costs.[5]

According to U.S. Department of Defense Instruction 5000.02, DCAPE is supposed to have timely access to any records and data from the U.S. DoD that "it considers necessary to review cost analyses and conduct the ICEs and cost analyses."[6]

To support this function, CAPE maintains several-cost reporting methods, including the Cost and Software Data Reporting (CSDR) system, which collects data directly from the contractors and stores that information in a CAPE repository. These data are the primary source of measuring contractor costs incurred from program development. DCAPE also gathers information through the Integrated Program Management Report and the Visibility and Management of Operating and Support Costs (VAMOSC) systems.[7] In addition to the above data, CAPE requires U.S. DoD component and service cost agency operation and support cost estimates developed at any time during the life cycle of a major weapon system. DCAPE also requires use of the Cost Analysis Requirements Description (CARD) and provides guidance on the content of the CARD. The CARD is supposed to be given to CAPE by the services prior to a planned Overarching

---

[5]   U.S. Department of Defense Instruction 5000.02, 2015, pp. 127–129.

[6]   U.S. Department of Defense Instruction 5000.02, 2015, p. 128.

[7]   According to U.S. Department of Defense Instruction 5000.02 (2015), VAMOSC data systems are managed by each military department and collect historical operating and support costs for major fielded weapon systems. DCAPE conducts annual reviews of VAMOSC systems to address data accessibility, completeness, timeliness, accuracy, and compliance with CAPE guidance. The annual reviews also assess the adequacy of each military department's funding and resources for its VAMOSC systems.

Integrated Product Team or equivalent staff-coordination-body review or U.S. DoD–component review.

The second major scrutiny role is program evaluation, which involves analysis and advice on matters relating to the planning and programming phases of the Planning, Programming, Budgeting, and Execution (PPBE) system. In fact, CAPE is the lead organisation for the programming phase,[8] with responsibility for assembling the Future Years Defense Program (FYDP), which covers five years of spending, and leading the annual program review process with the Under Secretary of Defense (Comptroller) for the Deputy Secretary of Defense. This gives CAPE wide latitude to review FYDP submissions by all of the U.S. DoD components to see if programs (new and old) are

- aligned with the most-current strategic needs and planning guidance
- cost-effective in light of current requirements and revised cost projections
- consistent with current U.S. DoD priorities, planning guidance, and fiscal guidance
- adequately funded
- consistent with programmed funding levels.

Additionally, as part of its program-evaluation role, CAPE provides a specific analysis of every MDAP (although it does not provide a complete portfolio review).[9] CAPE's multiple roles in the multistep, complex PPBE process are illustrated in the descriptions in red text in Figure 3.1.

In addition to cost assessment and program evaluation, CAPE has a third function: providing general analytic support to the secretary or deputy secretary. In this role, CAPE analyses other resource-allocation questions involving force structure, force deployments, reviews of high-

---

[8] U.S. Department of Defense Directive 7045.14, *The Planning, Programming, Budgeting, and Execution (PPBE) Process*, Washington, D.C.: Under Secretary of Defense (Comptroller), January 25, 2013.

[9] GAO, 2015.

Figure 3.1
CAPE's Role Within the PPBE System

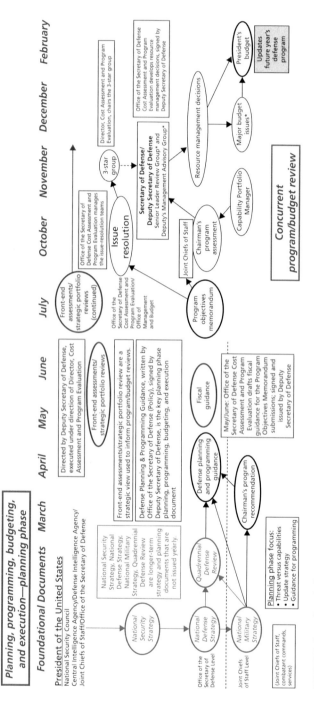

SOURCE: Based on information available on the DAU website and RAND knowledge.
NOTE: Red boxes reflecting CAPE's role were added by RAND. * = to be determined.

RAND RR1372-3.1

interest subsets of the U.S. DoD portfolio, workforce costs, overall organisational efficiencies, and other high-level defence issues. In these analyses, CAPE is often fulfilling a scrutiny-like role. For example, CAPE conducted reviews of overall force structure reduction options for U.S. Secretary of Defense Robert Gates for the 100 billion USD defence budget reduction in 2011 and again in response to the U.S. Budget Control Act ("Sequestration") for U.S. Secretary of Defense Chuck Hagel's Strategic Choices and Management Review in 2013.

To support its PPBE and analytic support roles, CAPE (working in conjunction with the Joint Staff) maintains the Joint Data Support office. This office functions as a repository of authoritative data on force structures both historic and projected though the FYDP and standardised reference information on threat capabilities and capacities, as well as a library of analytic models, such as Automated Behavioral Analysis Tool (ABAT), Extended Air Defense Simulation (EADSIM), Joint Integrated Contingency Model (JICM), and Synthetic Theater Operations Research Model (STORM). This repository is made available to authorised U.S. DoD users and supports a variety of uses, including campaign model and force structure analyses. CAPE also maintains the Data Warehouse[10] of programmatic data, including detailed FYDP data and DoD budget data for the current year and historic data; these data are made available to authorised DoD users and support various analyses of trends in U.S. DoD spending and program performance.

As can be seen in Figure 3.2, one important analytical function that CAPE performs in its third major role is to assist in the analysis of alternatives (AoA) for acquisition programs.[11]

---

[10] The Select and Native Programming (SNaP) data input system is part of the CAPE Data Warehouse.

[11] According to U.S. Department of Defense Instruction 5000.02: "The AoA assesses potential materiel solutions that could satisfy validated capability requirement(s) documented in the Initial Capabilities Document, and supports a decision on the most cost effective solution to meeting the validated capability requirement(s). In developing feasible alternatives, the AoA will identify a wide range of solutions that have a reasonable likelihood of providing the needed capability" (2015, p. 125).

Figure 3.2
DCAPE's Role Within the Capability Requirements Process and the Acquisition Process

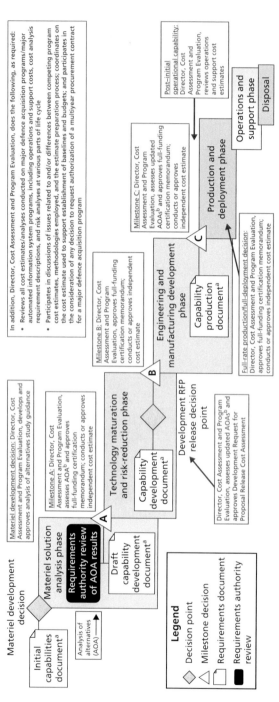

a Or equivalent approved/validated requirements document.
b Indicates that DCAPE only assesses AOAs for ACAT ID/IAM programs for the
U.S. DoD. For smaller programs, including ACAT IC, II, III, IV, and others, DCAPE will still review certain cost estimates even though it
does not perform them itself.

SOURCE: Modified from Department of Defense Instruction 5000.02, 2015, p. 5.
NOTES: Red boxes reflecting CAPE's roles were added by RAND, per guidance provided in Department of Defense Instruction 5000.02.

RAND RR1372-3.2

DCAPE develops and approves study guidance for the AoA for potential and designated Acquisition Category (ACAT) I and IA programs and for each joint military or business requirement for which the Chairman of the Joint Requirements Oversight Council (JROC) or the Investment Review Board is the validation authority. In developing the guidance, the DCAPE solicits the advice of other DoD officials and ensures that the guidance requires, at a minimum:

(1) Full consideration of possible tradeoffs among life-cycle cost, schedule, and performance objectives (including mandatory key performance parameters) for each alternative considered.

(2) An assessment of whether the joint military requirement can be met in a manner consistent with the cost and schedule objectives recommended by the JROC or other requirements validation authority.

(3) Consideration of affordability analysis results and affordability goals if established by the Milestone Decision Authority (MDA).[12]

In all three roles, CAPE has an advisory function. It never makes decisions, but rather presents analysis, recommendations, and alternative suggestions to the secretary, deputy secretary, and other senior leaders to inform their decisions.

When CAPE is considering programs to review, it considers risk, the capabilities that the program would provide, and how those capabilities fulfil the military's requirements. Analysts can break with popular opinion and question conventional wisdom, which makes it an interesting place for analysts to work. Figure 3.2 shows CAPE's involvement in the U.S. DoD's capability requirements process and the acquisition process; specific CAPE roles are highlighted in red text. The figure illustrates specific information requirements that CAPE is involved in at various major parts of these processes that inform decisionmakers.

---

[12] U.S. Department of Defense Instruction 5000.02, 2015, p. 125.

United Kingdom Ministry of Defence

The United Kingdom Ministry of Defence (UK MoD) has had a defence-specific scrutiny function for approximately 20 years. Information on the UK's scrutiny function was garnered from various sources, including published documents on UK processes, several discussions with a senior member of the UK MoD's scrutiny function, and a detailed presentation on the function provided by that official.

At the top of the UK MoD is the Defence Board, which "is [the] senior organisation in Defence and the main decision-makers for non-operational matters."[13] The Defence Board oversees three subcommittees: the Investment Approvals Committee (IAC), the Defence Audit Committee (DAC), and the People Committee. The key contestability function of scrutiny is carried out by a team of scrutineers on behalf of the IAC. Audit is a separate function that is undertaken by the DAC.[14] The Permanent Under Secretary serves as the departmental accounting officer for both the Defence Board and the DAC, and both the DAC and IAC are managed independently from the branches of the armed forces.

Outside the UK MoD, the Major Projects Authority has the responsibility to review all major projects in the UK government. When a defence project is reviewed by the Major Projects Authority, as many as three staff members will come to the project manager's office, review the project, and produce a report. That report is intended as advice to the project manager, and he or she may choose to share it with stakeholders.

The UK recently set up the Joint Requirements Oversight Committee (JROC), which is a subcommittee to the IAC. The JROC supports the IAC by considering requirement-setting in the early stages of the simplified acquisition cycle,[15] up to Main Gate approval. The JROC seeks to improve the ability of the IAC to understand and approve requirements for major programs, provide assurance that the

---

[13] UK MoD, *How Defence Works*, Version 4.1, London, September 30, 2014, p. 12.

[14] UK MoD, *Defence Audit Committee: Terms of Reference*, London, March 2013.

[15] This cycle is known as the concept, assessment, demonstration, manufacture, in-service, and disposal (CADMID) cycle.

requirement-setting process has been rigorous, and enhance oversight of those requirements that cross top-level budget boundaries as part of the Defence Capability Coherence function. A key outcome for the JROC is to ensure that requirements do not generate high levels of technical risk, which could carry potentially excessive costs and lengthy schedules, for proportionately low overall capability benefit.

Capability requirements are derived from capability gaps and are identified by the frontline command (e.g., air command). Each command has a separate budget with which it must align its requirements. These requirements would then form the capability part of the case seeking an investment decision.

Investment Approvals Committee and Scrutiny Team

The IAC reviews the largest defence investment proposals, including any investment requiring more than GBP 400 million (AUD 808 million).[16] Lower-value projects seeking investment are considered by the frontline commands within delegated limits, and these cases are also scrutinised by the central scrutiny team for all investments larger than GBP 100 million (AUD 202 million).[17]

The IAC is a four star–level committee, officially composed of the following senior leaders within the UK MoD:[18] Director General, Finance; Vice Chief of the Defence Staff; Chief Scientific Adviser; Chief of Defence Materiel;[19] Deputy Chief of the Defence Staff Military Capability; Director, Legal; and Director, Commercial.[20] The IAC

---

[16] Based on the average daily currency exchange rate between 1 December 2014 and 1 December 2015 of 2.0212 AUD to 1 GBP, from Oanda.com.

[17] Based on the average daily currency exchange rate between 1 December 2014 and 1 December 2015 of 2.0212 AUD to 1 GBP, from Oanda.com; also based on materials provided by the UK MoD.

[18] Based on materials provided by UK MoD.

[19] The Chief of Defence Materiel is responsible for, among other things, "providing equipment and logistic support to current operations[,] . . . delivering funded equipment acquisition and support outputs . . . [and] delivering projects to performance, time and cost targets." "Chief of Defence Materiel: Bernard Gray," web page, GOV.UK, undated.

[20] The Director, Commercial, is "responsible for the future direction and development of industrial strategy and the department's commercial relationships with industry." "Defence

members are the heads of the main administrative functions of the UK MoD, plus the top of the armed forces. Consequently, the IAC is a multidisciplinary group that can bring to bear different perspectives when considering the major decisions.

The IAC makes its investment evaluations at the initial and main gate decision points and as needed if a project is incurring extra risk, based on a portfolio of evidence, including the business case and the scrutiny report. The business case is provided by the project team or sponsor, and the scrutiny report is provided by the scrutiny team. The project team's business case will make an investment recommendation, while the primary role of the scrutiny function is to provide an independent, nonadvocate review of the evidence supporting the case (project) seeking an investment decision. The scrutiny reports are relatively short—on the order of ten pages—and focus on making a judgment about whether the evidence is sufficiently robust to support a new investment or a move to the next phase in a project.

The scrutiny function is divided into six groups, with about 60 members in total. These groups are operational analysis, technical, value for money, legal, affordability, and procurement strategy and commercial. Each group is led by a one-star equivalent who nominates a member of that group to form a team to participate in the development of any particular scrutiny report. The team is led by the member from either operational analysis or technical, depending on the topic. Scrutiny teams are staffed by civil servants and a few military, who are all experienced in their functional areas (e.g., operations analysis, program management, and commercial). While the project teams may need to bring in special assistance to develop their business cases, the scrutiny team does not.

The scrutiny team views itself as pragmatic, professional, and relevant, and it uses a "collegiate" approach, intended to assist the sponsor of the new investment with improving the analysis, without itself advocating for a specific decision about the investment.

---

and Armed Forces—Guidance: Ministry of Defence Commercial," web page, GOV.UK, December 12, 2012.

Once the IAC reviews the investment, it makes a decision as to whether the project can proceed. If the IAC approves, it then produces a ministerial submission to recommend approval by the ministries and by Her Majesty's Treasury. The UK MoD has delegation from the treasury to approve projects under GBP 400 million (AUD 808 million).[21] Once the investment for a project is approved, the capability sponsor for the project is responsible for project execution and monitoring over time.

To summarise, Figure 3.3 depicts the scrutiny process, including its relevant flow of activities and where the scrutiny function participates, based on RAND's understanding of how the process works; it is not an official UK MoD chart.

Germany

The German government has three main review offices. Specific military project scrutiny is conducted by the Bundesamtes für Ausrüstung, Informationstechnik und Nutzung der Bundeswehr (BAAINBw, translated to the Federal Office of Equipment, Information Technology and In-Service Support). Cost scrutiny is conducted by the National Audit Office. Reviews of general government expenditures are conducted by the Bundeskartellamt (BKartA).

The BKartA is an independent competition authority tasked with preventing monopolies and ensuring effective competition in both the public and private sectors. It is the responsibility of the Vergabekammern (in the BKartA) to review procurements and investments to ensure that the tenders were exercised legally.

The BKartA is embedded as a subordinate function of the Ministry of Economic Affairs and Energy. However, it operates independently to conduct and rule on investigations on abusive practices, mergers and acquisitions, and review of public award contracts.[22] The BKartA is made up of five divisions that serve distinct purposes, (1) Decision

---

[21] Based on the average daily currency exchange rate between 1 December 2014 and 1 December 2015 of 2.0212 AUD to 1 GBP, from Oanda.com.

[22] "The Bundeskartellamt," web page, Bundeskartellamt.de, undated.

Figure 3.3
UK Scrutiny Process (RAND Interpretation)

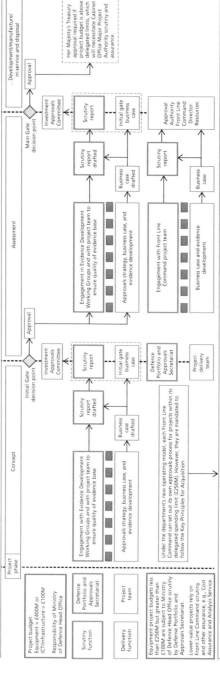

Her Majesty's Treasury has given the Ministry of Defence delegated authority for all projects of up to £600M (exceptions are for projects involving £400M of resource/personnel budgets and ICT projects where the delegation is £100M). Within the department approvals for non-ICT and infrastructure investments, up to £250M are delegated to the Front Line Commands (army, navy, air force, joint forces), unless the project is novel or contentious.

For projects which need Ministry of Defence Head Office approval, the scrutiny community (coordinated by the Defence Portfolio and Approvals Secretariat) tests the evidence behind the business case. The approving authority, e.g., the Investment Approvals Committee or Director of Resources, is given an assessment, in a scrutiny report, of the statements made in the business case on benefits, risks, and costs. This report tests the strength of the proposal, the soundness of future planning, and the validity of assertions in the business case and assesses whether they back up the recommendation effectively.

Capability sponsors and project-delivery teams involve the scrutiny community early in the approvals process. This allows the scrutiny team to build its understanding of (and confidence in) a project and assist the project team in determining appropriate levels of evidence required for each approval point.

Key Principles for Investment Approval

a. Fully justified and coherent with high-level defence strategy and Top Level Budget Command plans and integrated with other departmental and cross-government initiatives and projects

b. Designed to meet a capability or business requirement in a way that is appropriately flexible and adaptable to future military tasks

c. A cost-effective means of delivering optimised military capability or business benefits that offer through-life value for money

d. Affordable within existing and foreseeable future budget provision

e. Capable of being delivered on time and to cost, through effective acquisition management and commercial arrangements

f. Soundly based, with key risks to performance, cost, and time identified, and actions planned to monitor, manage, and mitigate those risks.

NOTE: ICT = Information and Communications Technology.

Division, (2) Litigation and Legal Division, (3) General Policy Division, (4) Federal Public Tribunals, and (5) Central Division.

The Decision Division is the ultimate decisionmaking authority of the BKartA and is composed of 12 decisionmaking bodies that are assigned an industry or sector. The Litigation and Legal and General Policy divisions act in advisory capacities in the legal and economic realms, and the Federal Public Tribunals works in public contracts. The Central Division (administrative) appears to be separate from the other functions. The BKartA issues annual reports in addition to topic or sector-specific inquiry reports, at its discretion.[23]

Within the German Ministry of Defence (MoD) and the BAAINBw (the BAAINBw is a separate administration supervised by the German MoD), there are four additional review and scrutiny mechanisms.

The first review mechanism exists within the MoD. There are nine General Directorate Offices, including the BAAINBw. The BAAINBw is responsible for the development, purchase, and evaluation of military equipment. Program and project cost evaluation is another major task of the program managers within the BAAINBw. The BAAINBw is a line-organised division that answers, through the chain of command, to the Defence Minister. The BAAINBw consists of multiple directorates (sea, air, land, etc.), and has existed in some form since 1950,[24] working on acquisitions for technologies necessary for defence. As with many of the other equipping and logistics agencies, the BAAINBw focuses on providing the support needed at a reasonable price point. The BAAINBw picks its own programs to review, and almost all programs are reviewed. It has only existed in its current form since 2012.

The second review mechanism involves the Rüstungsboards (Rü board), which was recently formed by the German MoD as the result of an external consulting agency recommending several organisational changes to increase scrutiny and the consolidation of the "controlling" function within the MoD. The function of the Rü board is to provide financial, technical, and schedule scrutiny for about 20 projects a year

[23] "Organisation," web page, Bundeskartellamt.de, undated.

[24] BAAINBw, home page, last updated November 11, 2015.

(those that are politically sensitive or over certain amounts). The Rü board is led by the State Secretary within the German MoD and contains fewer than ten people. There are several static positions on the board (including the state secretary, the head of the planning section, a three-star general representative from the German MoD, and the armaments division head), as well as a few positions that change based on the projects the board is reviewing (for example, the chief of staff and the program manager). The Rü board has an ad hoc staff and pulls in expertise as required.

The third review mechanism derives from a controlling function that exists at the directorate, BAAINBw, and German MoD levels. Every three or six months,[25] each project's various schedule, cost, and technical data are required to be added to a database. This database is then reviewed by the controlling function at the appropriate level (either the department head, the president of the BAAINBw, or the German MoD) for the size and criticality of the program. These controllers can only provide recommendations (up their chain of command—department, BAAINBw, German MoD) and do not have the authority to cancel a program.

The fourth review and scrutiny mechanism is the German parliament, which reviews all projects with a value greater than the EUR 25 million (AUD 37 million[26]) threshold, about 70–100 projects a year.

Initial project requirements are derived from the White Book (similar to a Quadrennial Defense Review or a White Paper). The trade-offs between capabilities and cost occur within the Planning Directorate, within the German MoD. The financing and planning capabilities focus on three periods: long term (five to ten years out), mid term (five years out), and short term (one to two years out). The German MoD also requires development money to be budgeted two years in advance, but just one year in advance for purchasing. As a result, integrated planning occurs throughout every year.

---

[25] RAND conducted an unofficial interview with a German official in the policy directorate who provided insight into the multiple scrutiny functions within the German MoD.

[26] Based on the average daily currency exchange rate between 1 December 2014 and 1 December 2015 of 1.4770 AUD to 1 EUR, from Oanda.com.

External analysts have been brought in to add to the constant review process. Portfolio review assessments cannot be made looking solely at the capabilities and developments of one country. As global defence becomes more connected, it is important to any decisionmaking process to understand fully the ally and Foreign Military Sales components of any country's defence portfolio, even though the review process for multinational projects may be additionally complex.

Canada

Canada has a stand-alone scrutiny and oversight panel within its defence structure. The Canadian Armed Forces is a joint institution with naval, air, and land components. Canada spent more than CAD 20 billion (AUD 21 billion[27]) on defence in 2015, with around CAD 35 million (AUD 37 million[28]) going to research and development. In February 2014, the government of Canada released its Defence Procurement Strategy. A key initiative of this strategy was to streamline all acquisition processes. Under the strategy, a special adviser, appointed by the Governor in Council, was tasked with creating a new, independent challenge function within the Canadian Department of National Defence (DND). This adviser drew on a range of sources, including the experience of key stakeholders from within and outside government, as well as Canada's recent experience with fighter and shipbuilding procurement processes. This process culminated in the establishment of the Independent Review Panel for Defence Acquisition (IRPDA) in June 2015. IRPDA serves as a third-party challenge to military acquisitions of CAD 100 million (AUD 105 million[29]) or more, as well as projects requiring Treasury Board approval and those identified for challenge by the Deputy Minister or Minister of National Defence. The five-member panel was designed around a strategy of early engagement, so that issues can be resolved at the front end of the procurement

---

[27] Based on the average daily currency exchange rate between 1 December 2014 and 1 December 2015 of 1.0454 AUD to 1 CAD, from Oanda.com.

[28] Based on the average daily currency exchange rate between 1 December 2014 and 1 December 2015 of 1.0454 AUD to 1 CAD, from Oanda.com.

[29] Based on the average daily currency exchange rate between 1 December 2014 and 1 December 2015 of 1.0454 AUD to 1 CAD, from Oanda.com.

process. IRPDA is supported by a small permanent staff—currently five people, but that number may grow to as many as seven.

RAND talked to a Canadian official to gain clarity on the contestability functions of IRPDA within the DND.

Requirements Versus Cost Analysis

Rather than performing a cost-review function, the panel analyses requirements. For example, a new order for some number of a new weapon system would not trigger a review of the price but rather an examination of why that specific number is required (cost reviews are left up to departmental officials and other governance processes). While each project is reviewed as a distinct endeavour, the panel acknowledges the importance of a holistic understanding of the armed forces.

Given that the scope and complexity of projects differ, as well as the small size of the staff of the IRPDA, these reviews do not go into the same level of detail as contestability or scrutiny groups with larger staffs. The reviews focus on established core areas of interest related to the project and its requirements, including the capability gap and quality of high-level mandatory requirements. The cabinet appoints all IRPDA members.

The office for IRPDA supporting staff provides the logistics and basic analytic needs of the panel. The senior staff members working for the IRPDA are largely generalists. Different staff members lead specific projects, depending on specialty. All staff are members of the senior cadre of civil servants, each with more than ten years of experience in their fields. While the current members are all former DND staff, this is not a requirement of the office, and future staff members might include others outside the department or external consultants. As of yet, the office has not felt the need to hire external consultants or contractors. The staff will rotate and will consist of senior analysts with professional maturity.

To help maintain the independence of the staff, their physical office is outside the DND. However, connections to the DND are vital because the department provides all the information needed for analysis, and the panel does not maintain its own databases.

The office analysts and the panel undertake reviews of programs at the end of the first two phases in the acquisition process: identification and options analysis.[30] The analysts are responsible for identifying material for the panel to review and interacting with the projects, as required, between the two end-of-phase reviews.

Multiple projects come before the IRPDA at the monthly meeting, and each is reviewed to see how it is meeting its high-level mandatory requirements and how it fulfils the capability gap and proposes to address it. To determine how well a project meets these requirements, the IRPDA reviews the project documentation and analysis provided by its office and then communicates with the project staff and key stakeholders. In some cases, the panel compares the project's initial requirements and its final business case, to determine whether any significant changes have occurred. The IRPDA does not review whether the estimated cost is appropriate; it only reviews capability requirements. Once a determination regarding the requirements is made, the panel secretary, who is also the executive director of the office, will prepare the IRPDA's advice to the minister on a project, which will then be sent to the minister in a memorandum from the IRPDA chair via the Deputy Minister of National Defence. Each memorandum is purely advisory.

While the metrics to judge the efficacy of the IRPDA are still under development, current factors include timely engagements, integrity of independence, and quality of advice.

Netherlands

The total annual budget managed by the Netherlands MoD was EUR 7.9 billion (AUD 11.7 billion[31]) in fiscal year 2014.[32] The Netherlands uses an internal challenge framework as its contestability func-

---

[30] The Canadian acquisition process has four phases: identification, options analysis, definition, and implementation.

[31] Based on the average daily currency exchange rate between 1 December 2014 and 1 December 2015 of 1.4770 AUD to 1 EUR, from Oanda.com.

[32] Netherlands Ministry of Defence, *Doing Business with the Netherlands Ministry of Defence*, The Hague, 2008, p. 13.

tion for defence acquisition but does not have a separate contestability office that oversees or manages these efforts.

The DMO of the Netherlands is one of the seven organisational elements of the MoD, alongside the four armed services, support command, and the Central Staff.[33] The DMO manages all major projects, defined as those costing EUR 5 million (AUD 7.4 million or more[34]). Nonstrategic projects—meaning ordinary operational logistics, real estate, etc.—are reported annually in the budget and the ministry's annual report.

The Netherlands defence contracts are banded by total estimated cost, not by type. Results of the contract are reported at each phase in the contract's life cycle. Reviews of these contracts are conducted internally but involve stakeholders from inside and outside the organisation. When outside expertise is required, subject-matter experts are consulted, as needed. If a particular project is desperately required, it may be reviewed by means of a "fast tracking" process.

The Netherlands MoD developed a four-phase process for significant acquisition projects with budgets of EUR 100 million (AUD 148 million[35]) or higher.[36] The process consists of (1) a requirement analysis, (2) a prefeasibility study, (3) a feasibility study, and (4) a final decision. Each phase ends in a formal review by a panel. The actors in the review panel include the Chief of Defence, who is the project sponsor; the user (i.e., the service for which the procured capability is intended); the procurement organisation (within the DMO); the Policy Directorate; the comptroller; and sometimes the personnel branch, depending on the topic. Each office is responsible for nominat-

---

[33] Netherlands Ministry of Defence, 2008.

[34] Based on the average daily currency exchange rate between 1 December 2014 and 1 December 2015 of 1.4770 AUD to 1 EUR, from Oanda.com and Netherlands Ministry of Defence, *Overview of the Defence Materiel Process*, The Hague, September 2007, p. 8.

[35] Based on the average daily currency exchange rate between 1 December 2014 and 1 December 2015 of 1.4770 AUD to 1 EUR, from Oanda.com.

[36] This monetary threshold has been stable for the past 13 years, but this is currently a subject of debate.

ing appropriate representatives for the review of each project. The participants are typically at the one-star-general officer level.

The original requirements for a program are defined by the Chief of Defence, who also maintains the yearly budget plan. Of the entire defence budget, about 20 percent is allocated toward investments. This investment plan is reviewed and updated at the beginning of the year, and then reviewed again midyear. As the projects progress through the four phases, the levels of risk and "risk cushion" built into each budget decrease.

Once a project has passed the requirements phase, the procurement office leads the remainder of the process. The next two phases, the prefeasibility study and the feasibility study, may be combined for less complicated procurements (i.e., "off the shelf" procurements). Among other criteria, these feasibility phases will look at the broad market; examine whether the performance required can be attained; and investigate sources, investment strategy, and partnership options. The feasibility phase also will compare these data with the established requirements and budget to conclude which solutions are practical. If the feasibility phase concludes that the requirements cannot be met within the budget, either the requirements or budget are amended by the Chief of Defence. The fourth and final phase is the decision phase. In this phase, potential providers are requested to enter their proposals. The decision phase entails final deliberation and the selection of a preferred proposal.

Panel members rigorously challenge various points of the study prior to moving on to the next phase. Any contested issues that cannot be resolved by the panel are included in the panel's final review report. RAND held discussions with a Dutch military official, working in an MoD policy organisation, to glean insight into contestability, and the official indicated that a distinct culture of honest criticism has been institutionalised into this review panel. Once the process is complete, the reviewing panel sends an abbreviated version of the final report to Parliament and the minister.

The Netherlands MoD has been relatively content with this process, which was introduced more than 30 years ago (as a result of a problematic procurement). However, a 2015 government-wide policy review recognised that the quality of acquisition decisions can be augmented by introducing a challenger. Ideally, the challenging function would manifest itself as a separate organisational entity, but the Netherlands does not have sufficient personnel to create such a department. As a result, the Netherlands MoD has considered strengthening the existing process by nominating the comptroller as the designated challenger for the panel. The comptroller is considered the most unbiased of the several panel participants, and it is considered inherent in the role to ask difficult questions during any potential expenditure review. Furthermore, as a result of the review, there is a more rigorous approach to cost-benefit analysis during the feasibility phase.

All acquisition data are housed within the procurement office of the defence ministry. The Chief of Defence relies heavily on these data when developing requirements and the associated budget for a particular project. Currently, the Netherlands sources most of its data from request-for-information submissions. Domestic and international research organisations are often used to validate figures. During the 2015 policy review, the maintenance of data was highlighted as an area requiring investment.

Risk

There is currently no formal risk framework in place. In each phase of the process, risks are identified, analysed, and—where possible—managed. In each project budget, a provision for risks is maintained. The amount of the provision varies from 5 to 10 percent, depending on the perceived risks of the acquisition. As is the case with many defence ministries, there is always a fine balance between developing capability in country and buying commercial off-the-shelf systems. The former is generally more expensive but can potentially deliver exact capabilities; the latter is often less expensive, but the capability delivered is already set.

Sweden

The Swedish Armed Forces is constitutionally an independent agency under the Swedish MoD, with a clear line between the armed forces and the Swedish MoD. The headquarters structures include the joint staff (Strategic Planning Office), the production staff, and the operational staff.

The armed forces rely on both strategic guidance and planning and long-term defence planning (LTDP) as ways to plan investments. They use strategic guidance and planning to identify and focus on major decisions, identify information needed for present decisions, provide a basis for prioritizing, give a foundation for logically based decisions, help to understand connections among different parts, and help manage uncertainties.[37] They use LTDP "to identify the need for strategic decisions and to propose alternative decisions and an assessment of these based on criteria."[38] Several plans are created based on the above methods: a 12-year logistics plan that includes procurement, a one- to three-year production plan, and an approximately five-year defence plan. The defence plan is included in the Parliament bill, and force structure is ultimately decided by Parliament.

Sweden understands itself as having a very constrained budget, with new major purchases requiring cutbacks in other areas. This means that new efforts are taken on only after the analysis described above. The Swedish MoD's small size also affects its ability to undertake preemptive analytical reviews, and it only performs reviews as a postmortem exercise for problematic programs.

Sweden does not have an independent contestability review function for defence and does not have a tradition of contestability reviews in advance of major investment decisions or on an ongoing basis throughout the weapon system life cycle. There are several agencies for general oversight purposes, and sometimes they get tasked to look at defence issues. One of these is the Swedish Competition Authority. In 2014, it reviewed a joint procurement of trucks by the Swedish

---

[37] Swedish Defence Research Agency, "Long Term Defence Planning—Purpose and Context," briefing, Stockholm, November 12, 2012; received from the ADoD via email on September 13, 2015.

[38] Swedish Defence Research Agency, 2012, p. 11.

Defence Materiel Administration and the Norwegian Defence Logistics Organisation,[39] and in 2013 reviewed several other cases.[40]

Denmark

Denmark employs several audit and oversight functions within its defence department. We address the general defence structure and these functions below.

Although Queen Margrethe II is the Commander-in-Chief of the Danish armed forces, as specified in the constitution, according to the Danish defence code, the Defence Minister acts as de facto Commander-in-Chief. Beneath the minister lies an agency named Defence Command Denmark, which consists of the Chief of Defence and the associated military leadership. Defence Command Denmark is the commanding authority of the armed forces.

The procurement of defence materiel and the oversight of such decisions are managed by the Danish Defence Acquisition and Logistics Organization (DALO), which falls under the defence ministry. The RAND team talked to a senior official of the DALO, who provided useful details about the organisation.

The DALO oversees a budget of DKK 7 billion (AUD 1.4 billion[41]), has about 2,400 personnel,[42] and manages the entire life cycle of the defence acquisition process in Denmark. All defence programs are sourced from the armed forces. The army, navy, and air force have distinct research and development centres in which they develop capabilities to meet specific military requirements. Once research ideas are presented to leaders within an armed force, they decide which designs will be presented to the DALO. At the appropriate time, the designated program manager will submit the investment plan and business case to the DALO. These meetings are rather formal in nature and can last up

---

[39] Konkurrensverket, *2014 Annual Report*, Stockholm, 2014.

[40] Konkurrensverket, *2013 Annual Report*, Stockholm, 2013.

[41] Based on the average daily currency exchange rate between 1 December 2014 and 1 December 2015 of 0.1981 AUD to 1 DKK, from Oanda.com.

[42] Værnsfælles Forsvarskommando, "The Danish Defence Acquisition and Logistics Organization," web page, undated.

to five hours. While this procedure applies to all military procurement and occurs weekly, the review process varies according to both project cost and strategic importance.

Large defence acquisitions are those that meet a threshold of approximately DKK 60 million (AUD 12 million[43]). Any projects at that cost level or higher must receive approval from the defence ministry, which can take up to four months. Once the DALO receives a project from a military branch, it will then revert to the ministry for review. After the minister grants approval for the specific program, the DALO can issue a formal tender of procurement to potential bidders. Acquisitions below the DKK 60 million (AUD 12 million[44]) mark are subject to internal DALO review. As mentioned, the comprehensiveness of the review is directly correlated to the cost and visibility of the project. For minor commercial off-the-shelf procurements, the reviewers of the acquisition can be the head of the business unit, department, or division within the military branch. Often the business case is rejected during review because of dissatisfaction with the quality of data. Once the decision has been made to proceed with the procurement, the program undergoes an internal audit to review all associated processes, procedures, and manuals. Once the audit is performed and all immediate risks are mitigated, approval is granted to sign the contract. Roughly 300 DALO employees are involved in this process.

From a macro level, the various defence institutions challenge each other during the procurement process. The DALO challenges the military branch that comes forward with the business case, and the ministry challenges the DALO for the procurements above DKK 60 million (AUD 12 million[45]). It is critical for the armed forces to come forward with acquisition ideas because they have the necessary expertise to determine Danish defence needs. The DALO man-

---

[43] Based on the average daily currency exchange rate between 1 December 2014 and 1 December 2015 of 0.1981 AUD to 1 DKK, from Oanda.com.

[44] Based on the average daily currency exchange rate between 1 December 2014 and 1 December 2015 of 0.1981 AUD to 1 DKK, from Oanda.com.

[45] Based on the average daily currency exchange rate between 1 December 2014 and 1 December 2015 of 0.1981 AUD to 1 DKK, from Oanda.com.

ages the thousands of proposals that are submitted (with a 25-percent approval rate), while keeping a keen eye on the budget and national defence strategy. The defence strategy is viewed with a 15-year planning horizon, and the military capability review is adjusted annually by an investment committee within the ministry. The ministry makes big budget decisions with political considerations in mind. There is an evident system of checks and balances in Danish acquisition that helps the country operate most effectively within its means.

With regard to data, the Systems Applications Products (SAP) system is utilised to manage all enterprise-level data related to projects, logistics, finance, and maintenance. In addition, the ministry uses a digital archive system that catalogues the agenda and content from every pertinent meeting. This effectively ensures compliance from all associated parties. The UK-developed PRINCE2 framework is used throughout the Danish defence system for integrated project-management processes. This provides high-level procedures on risk and is a valuable education for program managers.

New Zealand

New Zealand's contestability functions are embedded within their capability-development gate review process. New Zealand does, however, have a semiexternal panel to provide scrutiny at these gate reviews, as well as several internal methods of providing oversight.

The New Zealand MoD's Acquisition Division is responsible for the acquisition of the equipment used by the three services of New Zealand. Projects within the Acquisition Division are divided by service group. Defence expenditure is about NZD 2.3 billion (AUD 2.1 billion[46]) annually[47], and funds are to be spent where the need is greatest, according to internal rankings.[48] Total capital expenditures are

---

[46] Based on the average daily currency exchange rate between 1 December 2014 and 1 December 2015 of 0.9310 AUD to 1 NZD, from Oanda.com.

[47] New Zealand Defence Force, *The 2014–2015 Annual Report: For the Year Ended 30 June 2015*, Wellington: Ministry of Defence, 2015.

[48] New Zealand Defence Force, *Briefing for the Incoming Minister of Defence*, Wellington: Ministry of Defence, October 2014, p. 16.

expected to be about NZD 726 million (AUD 676 million[49]) in 2015.[50] In 2012, auditing and assessment of procurement was budgeted at approximately NZD 2 million (AUD 1.86 million[51]).[52] The Evaluation Division within the MoD is responsible for conducting independent assessments. There are few projects to evaluate annually, usually resulting in a little more than a dozen reports. Instead, success is measured by customer satisfaction and meeting budget requirements. This information is communicated in unclassified reports that are released annually.[53]

In July 2014, New Zealand's Parliament released the *Capability Management Framework* (CMF). This document explains the White Paper process, the capability and capital planning process, and the applicable laws. New Zealand's White Paper is a 25-year plan covering strategic requirements, and the capability plan sets out specific capabilities needed. The New Zealand MoD is careful to specify requirements, not solutions, because this approach enables greater transparency in investments (no predetermined solution is automatically procured). All government tenders are put on one website.[54]

Because the New Zealand Defence Force is small and joint, the CMF reflects some of the tough organisational changes that have been made to join capability and acquisition. RAND held a discussion with a senior member of the New Zealand Defence Force to better understand the capability development and contestability functions within the MoD.

As mentioned, New Zealand's major contestability function is embedded in its gate reviews. In this method, these reviews actually

---

[49] Based on the average daily currency exchange rate between 1 December 2014 and 1 December 2015 of 0.9310 AUD to 1 NZD, from Oanda.com.

[50] New Zealand Defence Force, 2014.

[51] Based on the average daily currency exchange rate between 1 December 2014 and 1 December 2015 of 0.9310 AUD to 1 NZD, from Oanda.com.

[52] New Zealand Defence Force, 2014, p. 27.

[53] New Zealand Defence Force, 2014, p. 26.

[54] The government tenders are put on the Government Electronic Tenders Service website (https://www.gets.govt.nz/ExternalIndex.htm). Access is open to all, after registration.

have the authority to stop a project. New Zealand has a five-gate review process, which includes a panel (independent of defence and including nonmilitary and ally members) review at each gate. Major capital projects (over several million dollars) must go through all five gates and are eventually signed off by the Prime Minister. All projects are reviewed and have a business case (which is presented to the review panel), but not all projects must go through all five gates (based on size and other criteria). Risks are noted in the business case for each project.

New Zealand employs ally assistance in the larger gate reviews. The State Services Division (a department that looks across the government) has a role is selecting panel members, but usually an ADoD and a UK MOD member will help with these reviews. The New Zealand MoD will facilitate the panel, but the panel's independence is maintained through its members' independence from New Zealand in general. Of course, there are different levels of review, and for smaller projects the review panel is at a much lower level and generally internal. Also, the operational arm of the New Zealand Defence Force is separate from the New Zealand MoD. The operational arm independently provides advice to the Secretary of Defence. The Secretary of Defence provides advice on capital acquisitions and thus prevents the New Zealand MoD from procuring above-required quantities (this happened on a previous vehicle acquisition) or capabilities that do not fit within the national portfolio. New Zealand has followed this five-gate and review process for the past five years, and its goal is to integrate effective sustainment decisions into the acquisition process.

The capability branch within the New Zealand MoD has a small standing office, but it has the ability to put together ad hoc teams for the hierarchies of review boards. Contestability outputs consist of a report that goes to the capability management board. Occasionally, capabilities and projects are rejected (even when they have a senior champion). New Zealand places a heavy emphasis on measurement. The Capability Plan (from the CMF and 25-year plan) is the document that drives decisions and forms the basis of measurement of whether a project will be a success. One of the largest problems New Zealand has seen with these types of review processes is when program decision control remains within separate services. Scrutiny is less effective when decisions can hide within a single service.

New Zealand uses a commercial requirements management tool to store all its program data and a commercial pricing model to capture all the aspects that go into costing the capability.

Other Government Organisations

U.S. Government Accountability Office

GAO is an independent agency, under the legislative branch, with two major missions.[55] The first is judicial, serving as investigator, judge, and jury in bid protests. The second is the auditing function. GAO employs about 3,000 people, all with a variety of specialties. Its fiscal year 2015 budget was USD 522 million (AUD 688 million[56]). To preserve its independence, it is led by the general comptroller, who is appointed for a single 15-year term.[57]

GAO engages in a variety of investigations and audits. In addition to bid protests, GAO will audit where requested, along the lines of about eight themes of programs that it decides on.[58] These themes include high-risk programs, programs of national importance, programs that are failing to perform, and other risk factors. In addition to bid protest reviews and audits, GAO often receives requests from congressional members and staffers for specific information, much like the Congressional Research Service (CRS). Over the course of these requests, it would be possible for congressional staffers to become acquainted with the personal biases of GAO staff, but they would not be easily able to shape the outcome of a requested piece of analysis by steering it to a particular staffer. Unlike the CRS, many of GAO's research activities

---

[55] GAO, home page, undated-b.

[56] Based on the average daily currency exchange rate between 1 December 2014 and 1 December 2015 of 1.3178 AUD to 1 USD from Oanda.com and GAO, undated-b.

[57] GAO, undated-b.

[58] RAND conducted an interview with a researcher who served on a congressionally mandated panel chaired by the comptroller general of GAO. The interviewee has extensive experience working with GAO and offered his professional judgment on the activities of GAO.

are based on preset audit activities. Over time, GAO has developed a reputation for independence and reliability.

This reputation is maintained through a careful focus on culture; there is a code of ethics established internally. A previous comptroller had a strong hand in implementing a wide array of internal performance metrics used to monitor progress, which both improved performance and raised morale. Morale is considered important because GAO can have a somewhat adversarial relationship to the rest of the government. GAO is often called the "congressional watchdog," and part of its mission is to ensure the government's accountability to U.S. citizens. In this respect, GAO's primary role is not to help the programs it is reviewing but rather to act as a mirror and simply reflect the findings of its audits.

Nonetheless, agencies typically cooperate during audits. GAO will contact a selected agency to request information. Once that contact is made, GAO will send in a local team (from one of its many field offices all over the United States) to collect the data and analyse it. Generally, all information is available to analysts. Once the data are analysed, GAO produces a report. GAO then releases its bid protest decision reports, budget reports, testimony, and letters to Congress. However, GAO has no legal authority to make changes. It is merely the reporter and auditor of facts.

The auditing portion of GAO is steeped in an accounting culture. A strong accounting culture means that employees are capable of performing many audits in exactly the same way. This method is not recorded or mandated, but rather is based on the historical performance of audits. This practice originates from the early years of GAO, when GAO was a part of the Department of the Treasury.[59] Originally, GAO only made sure that payments owed by the government were paid on time. Now, GAO focuses on accountability rather than merely accounting. This focus can include the need to examine the structure of a given project.

GAO is a strictly line-managed organisation; while employees may be cross-assigned to another team, they always answer directly to

---

[59] GAO, undated-b.

their supervisors. Analysts tend to concentrate on a particular specialty (a specific defence portfolio, health projects, etc.), and individual projects are audited and staffed with the analysts' specialties in mind.

Staff recruitment focuses on collecting skilled personnel with different backgrounds. This may include accountants, lawyers, or other specialists, with a variety of backgrounds, including many with advanced degrees. Despite having around 3,000 employees, projects are occasionally turned away because of a lack of personnel availability.

U.S. Government Agency

RAND held discussions with personnel within different contestability organisations in a large U.S. government agency that requested it not be identified by name. The agency's mission includes significant high-risk investment, and it has activities going on in multiple parts of the country. As with all U.S. agencies, this agency's total annual budget (sometimes referred to as the *top-line budget*) and some large pieces of the agency's operations are dictated by the U.S. Congress through appropriations laws. The U.S. Office of Management and Budget[60] (OMB) will dictate spending priorities for the agency at another level of granularity below the U.S. Congress, including significant decisions about the size, scope, and continued existence of major programs. In addition to budget and management decisions from Congress and OMB, the agency also has audit functions carried out by the agency's inspector general and separately by GAO. These audit functions may influence agency spending priorities by surfacing certain problems, but in contrast to OMB, the functions cannot directly dictate changes to the agency.

Within the boundaries set by Congress and OMB, the agency then engages in management and contestability efforts for strategic planning, setting priorities, and tracking the execution of work. The interviewees discussed contestability at the highest and lowest levels of the agency, and these functions have been in place for more than a decade. The agency has six organisations that perform contestability functions. The first three are funded by and report to the agency head,

---

[60] OMB reports to the President of the United States and is independent of other agencies.

separate from the operational units of the agency, with a fourth working not as an evaluation office but rather as a portfolio manager. The fifth and sixth are funded by the operational units and are described below. Only the first organisation has an oversight role, with some authority to dictate agency spending priorities. Also, these contestability organisations are all only institutionalised in agency policy, not law.

Contestability Organisations Funded by and Reporting to the Agency Head

First, at the top of the agency's management hierarchy, a contestability organisation—referred to here as the "Strategic Priorities Group" (SPG)—provides high-level oversight and scrutiny of large pieces of the agency's budget and work. The SPG is relatively small, is managed by the agency head, is funded separately from the operational parts of the agency, and contains a collection of technical and management expertise. The SPG surveys the long-term expectations for agency work using internal information and internal and external subject-matter experts, and it creates a top-level strategic plan that guides the budget and investments for the next few decades. This top-level strategic plan is used by the SPG to provide oversight of and scrutinise the proposed plans of the operational units, and ultimately make recommendations to the agency head about what programs should proceed and the sequence in which they should proceed. The SPG has sufficient staff and access to subject-matter experts to consider agency priorities from the perspectives of technical feasibility, managerial feasibility (e.g., staff and facilities), and cost. A key part of the SPG's oversight and scrutiny roles is a cost and risk formula that enables the organisation to quantify the merits of proposed work. In addition, the SPG may hire outside experts to help review certain significant events in a combination audit and scrutiny role—e.g., for accidents during agency operations.

Second, separate from the first organisation, but also at the top of the agency's hierarchy, a contestability organisation performs scrutiny of agency operations by conducting independent cost estimates— referred to here as the "Independent Cost Group" (ICG). The ICG is also relatively small, managed by the agency head, and funded separately from other agency operations. All agency programs are required to provide data to the ICG at multiple points in a program's life cycle to validate and help improve cost estimates across the agency. The ICG's independent cost estimate is presented to agency management for comparison with the cost estimate provided by the operational unit. The ICG also has access to an agency-wide database of the historic costs of some programs to help build parametric models for cost estimating.

Third, a contestability organisation—referred to here as the "Program Tracking Group" (PTG)—performs oversight, due diligence, and auditing at the lowest level of agency operations to track the progress of work and ensure compliance with agency procedures for review and approval. As with the first two organisations described here, the PTG is relatively small, managed by the agency head, and funded separately from other agency operations. The PTG tracks the execution of agency activities, with a particular focus on cost, schedule, and progress toward major milestones. If the PTG notices negative trends happening, it will notify agency management and recommend further review.

Fourth, a portfolio management office constantly examines the complete technical portfolio of the agency. This full portfolio is briefed to the agency's administration twice a year. This serves not as an oversight or due diligence function but as a way to maintain the agency's understanding of its own capabilities in a holistic fashion.

Contestability Organisations Funded by the Operational Units

Fifth, a contestability organisation performs scrutiny of proposed agency activities on an ad hoc, or as-requested, basis—referred to here as the "Independent Assessment Group" (IAG). The IAG is small and funded by the operational units, but it is outside the management chain of those operational units, which allows it to maintain a greater level of independence for its assessments. The IAG has built an internal reputation for itself—and top-level management support—sufficient to entice the operational units to use program funds to pay for an independent assessment. In other words, the IAG operates like internal agency consultants.

The sixth contestability element is that all agency operational units are required to hire outside consultants to generate cost estimates for large activities. These outside cost estimates are required by agency management but are funded by the operational units—i.e., they are built into the budgets of the discrete agency activities. Part of the reason for using outside experts is that this agency does not have sufficient internal analysts to perform this function.

The scope and scale of the agency's operations and level of risk in its investments are large enough to warrant multiple contestability organisations at different levels of agency management, but they all appear intended to contribute toward two agency goals: efficient use of agency funds and predictable progress toward program outcomes. The interviewees believe that the contestability organisations described here serve important roles and are helping to improve how the agency sets priorities and executes operations. This agency is no different from any other that is responsible for managing the uncertainty of advanced research and highly technical activities, and it has incorporated contestability to help address the problems that are inherent in the work.

National Audit Office

The UK MoD is also subject to investigations by the UK government-wide auditing and administrative investigations department, NAO. (Note that this description is based on publicly available documents rather than direct data collection from an employee of that organisation.) The NAO is responsible for auditing public expenditures to make sure that money is used legally and effectively.[61] While GAO adjudicates contract bid protests, these types of disputes in the UK go to another department outside NAO. Much like GAO, the NAO's head, the comptroller, answers directly to the Parliament.[62] The NAO was officially established in 1983, but an audit department has existed in some form since the 19th century. Approximately 800 employees work at the NAO, including statisticians, economists, and accountants.[63] Employees also have areas of specialty, such as health care, military expenditures, and education. The NAO has a clear chain of command. However, it is managed by a board, because of previous questions of

---

[61] NAO, home page, undated.

[62] NAO, undated.

[63] NAO, undated.

accountability.[64] The NAO is ultimately an advisory office that does not have the authority to stop a project or investment, but it can influence budget decisions through investigations and reports. These reports are made directly to the Parliament—with about 60 "value for money" investigations occurring annually[65]—the NAO reports are most often made public. A source of internal government conflict comes about in the drafting of these reports; the subject of investigations is given the opportunity to review audits (to ensure factual accuracy) before the final report is sent to Parliament.[66]

Private Companies

International Shipbuilding and Transportation Firms

RAND performed open-source research and talked to an executive of an international shipbuilding and transportation firm to gain insight into contestability within the private sector. The firm is a comprehensive solution provider within the shipbuilding industry and provides design, engineering, repair, and project-management services for a wide variety of ships. The firm's expertise includes patrol, defence, and supply vessels, as well as rigs, container ships, and tankers. The interviewee described the valuable insights that red teams provide during independent assessments. Although the firm may employ other review panels or techniques, the red-teaming exercises were the primary subject of the interview.

During a typical red-team exercise, the firm will invite representatives from a host of government and nongovernment entities to provide expert advice on a wide array of shipbuilding and industrial matters. The firm also has a strategic partnership with a peer in the industry and is able to draw on its expertise for red-teaming as well. Red teams

---

[64] P. Dunleavy, Christopher Gilson, Simon Bastow, and Jane Tinkler, *The National Audit Office, the Public Accounts Committee and the Risk Landscape in UK Public Policy*, London: London School of Economics, Risk & Regulation Advisory Council, 2009.

[65] Dunleavy et al., 2009.

[66] Dunleavy et al., 2009.

typically consist of five assessors from distinct organisations, to promote diversity of thought. This firm has found that it is critical to have red-team representatives from dissimilar backgrounds because they all bring unique expertise from their respective fields. Red-team sessions are conducted approximately three times per year and cost around AUD 175,000[67] on an annual basis.

One example of the successful use of a red team was after a recent and lengthy bidding process. The firm was awarded a long-term shipbuilding contract, and the strong certainty of business had the prospect of completely changing the dynamic of the organisation. The existing shipyards did not have the capacity to build the required number of ships at a reasonable cost. With a potentially substantial expenditure decision to be made, it was necessary for the firm to conduct a significant review of all aspects of the investment. For this reason, the firm used a red team during the review and eventually implemented some of the red team's recommendations by transforming its infrastructure with an investment of more than AUD 350 million, aimed mostly at the modernization of its shipyards. Our interviewee stated that there are two major factors when deciding whether to invest: (1) certainty from a customer base, and (2) a return on that investment within two to three years.

The firm has a unique structure: A corporate-level individual who is closely involved in the review process and greatly appreciates the red teams has oversight of activities. This leader will usually spend five hours with the red team during the final day of the session to accurately grasp its major considerations. Once the session is completed, the organisation receives, on average, three to four exceptionally useful recommendations that are executed and ultimately affect strategy. A second benefit of the red-team process, in addition to the recommendations, is that the preparation required to support a red team often illuminates previously hidden or not fully mitigated risks. The personnel specifying requirements often have no insight into their respec-

---

[67] All currency conversions for private companies were translated between the native currency and AUD, based on the average daily currency exchange rate between 1 December 2014 and 1 December 2015, as listed on Oanda.com.

tive costs. This information becomes important when conducting cost-capability trade-offs. The scrupulous review associated with red-team visits has fostered improvement within the organisation.

To maintain the unbiased nature of red teams, the firm regularly assesses the teams to determine whether any improvement to personnel can be made. There are times when red-team representatives disagree with each other on various decisions. In the event that this happens, senior leadership meets and determines which argument best supports the firm's best interests.

A second unique aspect of this firm is the strategic partnership it maintains with an industry peer. This strategic partner provides standards, procedures, and data that would be costly for the firm to develop and maintain on its own.

Risk

The firm manages risk by placing a fiscal value on specific contingencies on the proposed project. In the example of the modernization of a shipyard, or the building of a new ship, it was vital to establish a new framework for contingencies. In one example, existing standards for the welding of steel, pipe construction, and material sourcing were considered invalid and had to be recalculated. Some of this risk was discovered by the red team, and some of it was mitigated through the use of the strategic partnership. In essence, periodic red-team or external reviews of any high-value decision, consisting of several budget line items, help this firm recognise faulty assumptions, as well as previously unidentified costs and risks.

International Conglomerate Firms

RAND conducted an additional private-sector interview with a senior executive of an international conglomerate in a variety of industrial sectors, including transportation, to gain a better understanding of contestability functions that exist within private-sector companies. The findings below are informed by both the literature available on this company and information provided by the interviewee, who plays an active role in a scrutiny function. This firm has multiple core businesses and has experienced significant growth in the past ten years, particularly in the shipping part of its portfolio. As a result of the significant growth, the company has recently chosen a back-to-basics approach in its decisionmaking regarding investments.

This company has multiple formal contestability functions at multiple levels within the company: scrutiny (group strategy team and risk-management team), due diligence (finance team), and audit (internal audit team). Actual investment decisions are not made by these teams. Decisions are made by an executive-level CFO, CEO, and boards of directors, along with additional CEOs and boards of directors within the individual business units. The interview focused on the scrutiny function, or "independent review" function, within the company; however, the firm also employs due diligence before investment decisions and auditing, as part of the normal business operations. The particular scrutiny contestability functions (the group strategy team and the risk-management team) were formally established ten years ago and have evolved organically during the past 100 years. These functions have always existed informally as best business practices over time.

The company is organised into five core businesses, or "business units," that are independent of each other; however, these units compete for the same pool of corporate-level investment funds each year, of around AUD 13 billion. Typical investments can be maintenance of major assets, improvements in technology, or investments to increase market share. The company has a very specific investment strategy in place and makes decisions based on market conditions within each industry, consumer needs, and lowering costs, but tries to avoid "fancy" add-on services. Each business unit comes up with a yearly plan for how to achieve the above goals in its particular industry and then competes among units for funding in the capital-allocation and portfolio-allocation phase.

Investment decisions are made at both the business-unit level and at the centralised or executive corporate level. Lower dollar-value decisions (approximately AUD 25–65 million) are scrutinised by staff whose role is to understand strategy and risk at the business-unit level. Each business unit has its own management team, board of directors, and manager or CEO, which also actively take part in independent scrutiny of these lower-level decisions. At the corporate or enterprise level, the units have a group strategy team, risk-management team, internal audit team, and financial team. Larger investments must be elevated to an executive board of directors (which includes the group's CEO and CFO and the CEOs of four of its core business areas). The board of directors within each unit has a supervisory role, while the executive board is responsible for overall operations, financial results, and business growth.

The enterprise-level contestability function of the group strategy team strives to understand whether investment decisions are appropriate, along with potential alternatives. The enterprise-level risk-management team focuses on identifying and understanding risks associated with investment decisions. Both challenge and question business-unit plans. Within the entire organisation, the business units know, recognise, and understand the role of these independent functions, and the culture of the company supports these functions.

The company, in general, has a very lean corporate centre. Business units have a lot of control over their decisions, while the corporate functions are in place to add value. This can be seen in the enterprise-level group strategy team, which is only staffed with six to eight people and is typically very busy. This team is composed of mostly senior-level staff, with some junior staff members in support roles. Senior members have a lot of experience in the particular industry they are overseeing (e.g., a senior member may have many years of experience in the oil industry and can therefore easily understand industry trends and best practices). Given the lean organisation, the scrutiny function of this company relies on and takes advantage of external expertise that will provide value to the scrutiny of its investment decisions. The heads of the group strategy and risk-management teams report to the overall group CEO but do not sit higher in the organisation than the CEOs of the business units. These positions require significant subject-matter expertise. The length of time in the position varies based on internal company needs.

When performing the contestability function, the group strategy team, in particular, uses a mix of analysis and data provided by the business unit and its own independent analysis. The information and analysis follow a standard template, but not all the output fits precisely in this template, so the output takes on additional forms. The output is deliberately stored and reviewed every few years by both the group strategy team and the finance team to measure whether the investment decisions were successful. "Success" is measured by whether the investment does what it originally promised, whether the rate of return is acceptable given market conditions, and whether the business case was implemented. This company, as with most others, has challenges in managing the large volume of data that are collected throughout the various parts of the company.

Risk

The risk-management team at the corporate level has six to eight staff members, as well as additional staff in the business units who also analyse risk. The enterprise-level risk-management team is proactive in regards to risk management. It conducts a yearly review of risk within the company. Risks are identified by and provided to the enterprise-level team by business units, external sources, the executive management team, the strategy team, and the financial team. External analysts, based on their subject-matter expertise, are regularly consulted for information about applicable business-related areas (e.g., the oil outlook, other institutions that monitor risk, and economic forecasts).

This company tries to balance cost and capability trade-offs in every decision. Given the main business areas of this company, it is particularly keen on minimizing safety risks to personnel and will incur extra cost to protect its personnel. It also has clear lines for accountability within the company; the staff understands the roles of the different contestability functions and understands how the various functions interact.

International Security and Aerospace Firms

The commercial defence industry provides an interesting perspective on contestability within the military framework. A senior executive in the private defence sector described the internal oversight and audit functions used at the firm. This contractor provides weapon system, information technology, and aerospace solutions for various defence departments throughout the world.

The first oversight function pertains to market and portfolio investments and is named the Strategic Planning Group. The group examines the projected investment plan and ensures alignment with the firm's macro business strategy. Investment decisions are necessary throughout all strata of the organisation, and oversight affects both the magnitude and timing of decisionmaking. The staffing of the review panel depends on the scale of the decision, but the lead investigator must work in a separate department of the company. Market investment recommendations typically begin from the lower levels of the corporate hierarchy. Engineers will provide a list of requested technologies to their respective program managers, who will in turn communicate the recommendations up the management chain. Once an investment decision needs to be made, the review panel will be formed. Portfolio investment decisions originate from the higher echelons of the organisation but are still subject to the same review process. A challenge encountered with portfolio shaping in the defence sector is long-term investment. There is a tendency to favour near-term investment because the pressure to produce financial gain is immediate. Investing in a ten-year future may be the best solution, but it would not benefit the responsible individuals in the short term. The firm is aware of this quandary and has allocated approximately 20 percent of the total budget toward long-term investments.

A critical note to take into consideration is the motivational difference between the private and public sector. The commercial space is measured exclusively by financial results, while government success is subjectively measured by positive public benefits and by a lack of negative outcomes. Sales will diminish if investment is prioritised incorrectly. Consequently, there is an optimization function that is ubiquitous within the private sector. From the perspective of the board, this rationale negates the need for an external oversight function.

The audit function relates to program review and is called the program-excellence function. The program-excellence team is an internal function and reports directly to the vice president. Not dissimilar to CAPE, the program-excellence function reviews the status of a particular program through an internal quality-assurance framework. The function is staffed by senior individuals with backgrounds in audit and program management, some with graduate degrees. The size of program-excellence teams varies based on project size, ranging from four to 20 personnel. A senior figure from a different department is assigned as lead and then recruits necessary expertise as needed. External analysts have occasionally been employed, most typically for high-profile program reviews. Before a project bid is submitted, the program-excellence team must perform an independent, nonadvocate review of the proposal. This process is standard throughout the organisation and occurs before submission to executive leadership. Once the program has commenced, the function performs multiple reviews throughout the life span of the project, including program submittal and execution. The final product of the review consists of a presentation of the findings and recommendations. Concerning funding, the organisation pays for review during the bid-submittal process. The remaining reviews are built into the total program cost as overhead.

With respect to risk, each program is required to implement a risk-management program. The identified risks are entered into a central database and then assigned an associated mitigation program. Risk calculation is measured by probability and effect, from a scale of one to three. A waterfall diagram is then employed to further analyse the data. Any identified risks should be mitigated if possible because if an assumption does not hold, it is essential to have an alternative in place. In truth, the tool will only be effective if the assumptions and risks are well thought out. In that sense, it is vital to train risk analysts not only how to complete the database but also in the critical-thinking process involved. The program-excellence function uses this framework for their objective analysis, so the assumptions used are very important.

Summary

As is evident from the case studies described in this chapter, there is significant variation in how different organisations approach contestability and review functions. We provide a summary overview in Table 3.1. It lists the defence organisations by total defence budget, including an entry for Australia to show where it would stand in this group. The UK NAO and two U.S. government agencies are next, followed by the three commercial firms.

   In the next chapter, we describe some summary insights on different models, as well as the varying inputs to the supporting aspects of contestability models (e.g., desired objective, size, funding, and organisational location).

Table 3.1
Contestability Function Summary Characteristics

Organisation	Country Defence Budgets (FY 2014 AUD, Millions)[a]	Main Contestability Function Examined	History	Number of Staff	Stand-Alone Function	Direct Report to Decision-maker	Decision-making Authority	Type of Engagement	Outputs/Recipients
U.S. (CAPE)	803,744	Scrutiny	Originated 1961, current structure dates from 2009	160	Yes	Yes	No	Constant, ongoing	Mostly internal; Secretary of Defense/ Department of the Secretary of Defense
UK (UK MoD scrutiny function)	79,704	Scrutiny	20 years old	60, with 15 for technical and operational analysis scrutiny	Yes	Yes	No	Constant, ongoing	Internal; scrutiny; report to Investment Approvals Committee
Germany (BAAINBw and Rü Board)	61,218	Scrutiny	~2014	Fewer than 10	No (chairman and ad hoc board)	Yes	No	Constant, ongoing	Annual report to Parliament; assessment to Chairman of Rü Board
Australia	33,487								

Table 3.1—Continued

Organisation	Country Defence Budgets (FY 2014 AUD, Millions)[a]	Main Contestability Function Examined	History	Number of Staff	Stand-Alone Function	Direct Report to Decision-maker	Decision-making Authority	Type of Engagement	Outputs/ Recipients
Canada (IRPDA)	22,998	Scrutiny, oversight	Established in June 2015 (still in develop-ment)	10	Yes (formal panel and stand-alone supporting office)	Yes	No	Constant, ongoing	Memorandum from the panel chair via the deputy minister of national defence to the minister
Netherlands (DMO)	13,292	Scrutiny, oversight	Evolved from a process started in 1980s	No dedicated staff; other duty as assigned	No (ad hoc panel with potential future challenge function)	Yes	No	End of each major phase	Final report to Parliament and the minister
Sweden	8,662	Oversight	N/A	No dedicated staff; other duty as assigned	No (review as part of strategic and long-term planning)	No	No	Constant, ongoing	Internal and external; minister for defence and Riksdag
Denmark (DALO)	5,874	Audit, oversight	Established in 2006 as DALO	No dedicated staff; other duty as assigned	No (review as part of capability-development process)	Yes	No	As needed for large investments	Internal; defence minister

Table 3.1—Continued

Organisation	Country Defence Budgets (FY 2014 AUD, Millions)[a]	Main Contestability Function Examined	History	Number of Staff	Stand-Alone Function	Direct Report to Decision-maker	Decision-making Authority	Type of Engagement	Outputs/Recipients
New Zealand (Gate Review Panel)	3,175	Scrutiny, oversight	~2010	No dedicated staff; other duty as assigned; external assistance	No (ad hoc review panel)	Yes	Yes	Panel convenes at gate reviews	Internal; prime minister/ secretary of defence

Nondefence Organisations (public agencies and commercial firms)

Organisation	FY 2014 Revenue or Sales, AUD, Millions	Main Contestability Function Examined	History	Number of Staff	Stand-Alone Function	Direct Report to Decision-maker	Decision-making Authority	Type of Engagement	Outputs/Recipients
UK (NAO)	N/A	Audit	Established in 1983, in function long before	800	Yes	Yes	No	Constant, ongoing	Internal/ external; reports to Parliament
U.S. (large agency's ICG)	N/A	Scrutiny, oversight	2005 (in current form)	No information	Yes	Yes	No	Constant, ongoing	Internal; head of agency
U.S. (GAO)	N/A	Audit	Founded in 1921	3,000	Yes (stand-alone agency)	Yes	No	Constant, ongoing	Internal, but mostly external for public; Congress

Table 3.1—Continued

Organisation	FY 2014 Revenue or Sales, AUD, Millions	Main Contestability Function Examined	History	Number of Staff	Stand-Alone Function	Direct Report to Decision-maker	Decision-making Authority	Type of Engagement	Outputs/Recipients
International shipbuilding and transportation firm	1,307 (sales)	Scrutiny	No information	5 people from various outside firms	No (ad hoc—red-team sessions are about 3 times per year)	Yes	No	About 3 times per year	Report to senior leadership
International security and aerospace firm	60,000+ (net sales)	Oversight, audit	No information	Depends on scale of the decision	No (ad hoc strategic panel and program excellence team)	Yes	No	As needed for investments	Report to executive vice president
International conglomerate firm	60,000+ (revenue)	Scrutiny, audit, due diligence	8–10 years old (existed organically back to 1904)	Strategy (6–8); risk (6–8); finance/accounting (40–50); internal audit (10–15)	Yes (multiple enterprise level); ad hoc (lower level)	Yes	No	Constant, ongoing	Investment review report to CEO, CFO, board of directors

a Based on average daily currency exchange rate between 1 December 2014 and 1 December 2015 of 1.3178 AUD to 1 USD, from Oanda.com.

SOURCE: Military-spending figures from Stockholm International Peace Research Institute, SIPRI Military Expenditure Database, undated..

Summary Insights

The case studies in Chapter Three revealed that different organisations take a wide variety of approaches to implementing and conducting contestability functions. Each organisation has developed unique approaches to putting in place checks and balances that govern its decisions connected with large capital expenditures. That said, some clear common themes carry across many of these case studies, which we offer here in Chapter Four.

Along with these themes, several issues became clear over the course of the discussions. There was no single role that all of our interviewees held, and not all of our contacts had equal visibility into the processes at their organisations. These factors limited our ability to make exact comparisons across organisations.

Furthermore, we found that the defence contestability functions do not have clear metrics to measure their success. Their goal is to improve decisionmaking, not to approve or cancel programs. In no case for government contestability did we find simple measures of performance of the function that could be linked to the strategic outcomes desired by the organisation as a whole. In the private sector, the contestability functions did try and link outcomes more often to quantitative metrics (e.g., rate of return on investment), but these metrics do not transfer to public-sector organisations. Consequently, we cannot offer a menu of contestability metrics. A much bigger implication is that without clear metrics linking the structure and process of contestability functions to system outcomes, it is impossible to claim that particular activities fall under the rubric of best practice. We cannot definitively claim, based on this research, that a particular model of contestability is the best or most effective. Rather, each contestability approach appears to be contingent on the local defence ecosystem. The ADoD can proceed with the knowledge that there are several reasonable ways to insert contestability into its processes, as long as it adopts a function with appropriate independence, authority, and resources.

The History of the Contestability Functions

A review of the various contestability functions for this analysis showed that there were multiple ways that these functions came into existence. In some situations, the current function has evolved organically over time, as is the case with the U.S. DoD CAPE scrutiny function, which originated in 1961 as a formal office but has had multiple name changes since. The UK's scrutiny team was formalised in the 1990s, but the function was less formal before then. Another large U.S. agency also mentioned several functions that have also existed in some form over the years.

Also, several defence organisations have conducted or commissioned reviews that have led to the standing up of a new organisation or the desire to strengthen the existing approach. We found that several countries (Canada, Germany, and the Netherlands) recently held policy reviews recommending that each develop a new, or strengthen the existing, contestability function. Canada's response was to develop the formal IRPDA office, which is the stand-alone scrutiny and oversight panel described in Chapter Three. The Netherlands, on the other hand, has not to this point instantiated a formal office. Concerns about the resource burdens associated with adding a second round of review after the existing joint approach led Dutch defence leaders to avoid developing a new separate structure in spite of the recommendation. Instead, those defence leaders are considering strengthening the comptroller role and using the enhanced comptroller functions to strengthen the existing "challenge" function. Germany has regular reviews of its acquisition system that are usually conducted by a commercial consulting company and evolved its scrutiny function to the current version after the last review, which includes the Rü board.

In a few cases, poor acquisition outcomes (issues with cost, performance, or schedule) have led to the addition of a review function. For instance, the existing review structure in the Netherlands was stood up after a problematic procurement in the 1980s. New Zealand had a few troubled acquisition programs, including one that suffered because of a lack of clarity among participants as to their various roles and missions, and one where sustainment costs were not taken into full consideration during the acquisition process and costs ended up being unexpectedly high and burdensome.

Contestability reviews were not the only mechanism mentioned by interviewees to obtain best value. Competition for new weapon systems was mentioned as a way of ensuring a reasonable price. However, one interviewee did note that competition was easier to incorporate into acquisition when requirements focus on meeting specific mission needs rather than ensuring technological overmatch. This meant that acquiring existing weapon systems was likely to be an available option, which increases the possibility of competition. Defence organisations focused on buying technologies that advanced the state of the art were more likely to have to engage in high-risk development programs, with greater likelihood of failure. Competition cannot replace more-formal and more-structured contestability reviews in those cases.

Benefits and Caveats

The individuals (both in the public and private sectors) we talked to typically expressed the need for their organisations to allocate available resources carefully. In that context, they understood that contestability reviews (in all their forms) could improve that allocation by ensuring an independent look at decisionmaking. However, there was no single approach to organising, performing, or managing these reviews in the main contestability functions that were focused on by interviewees.

While our interviewees generally agreed that contestability functions improved decisionmaking, they also made it clear that the function does not provide a magic bullet for avoiding acquisition challenges. Cost growth on weapon systems is very common in the United States, in spite of the strength and size of CAPE. Another note of caution is that contestability that focuses on reviewing recommendations *after* they have been made (which is the most common approach) may be too late to fix them. However, the existence of a contestability function may lead the initial decisionmakers to perform more-careful analyses to ensure that their results hold up to scrutiny. It may also be that cost growth would have been higher without the careful review. Given the fact that a contestability function offers a system of checks and balances, we would assert that contestability itself is a best practice, although there are different ways of getting there and effects are often hard to measure.

In this research, there were cases where effective reviews of programs resulted in real change—including program cancellations. Others acted only in advisory roles, so they are unable to completely push back against powerful military service or political interests. We learned of specific challenges related to cases where strong political interests affected decisionmaking. These may be states or provinces in a federal system[1] or countries with strong defence industries where national interests in supporting an industrial base, a technical capability, or just jobs in general may be at odds with what the contestability function may suggest is a more cost-effective approach. In democracies, the elected officials are the ultimate decisionmakers and exercise that authority through appointed officials. It is the role of contestability functions to inform the appropriate officials so they can make informed decisions. It is the job of decisionmakers to also consider political, economic, social, and other considerations that are outside the purview of a defence contestability function, which may well override the recommendations made by the contestability function. So success of a defence contestability function cannot be measured solely by how many of its recommendations are implemented.

---

[1] The U.S. DoD has faced challenges moving capabilities from the National Guard to the active components because of guard supporters at the state level and in Congress.

We also note that our European interviewees frequently referred to European Union regulations on defence procurement that shaped their overall acquisition approaches.

Insights on Aspects of Contestability

The literature review summarised in Chapter Two found a number of aspects of contestability that could vary across approaches. These contributed to our discussion protocol, and we describe some summary insights below.

Focus of the Function

There was some variation in the focus areas of contestability organisations. Military contestability functions most commonly focused on significant procurement decisions, with one (U.S. CAPE) also engaging in analyses of larger resourcing allocation issues, such as force structure, workforce costs, and organisational efficiencies. No organisation reviewed all the possible decisions. Rather, there was typically some monetary or budget threshold necessary for formal reviews by the contestability function. For example, higher-dollar programs in terms of a country's defence portfolio were more likely to be reviewed by a formal scrutiny function, while lower-dollar ones were reviewed using less formal contestability functions, such as oversight at lower levels in the organisation. A few interviewees expressed some frustration that these thresholds had been set some years back and had never been adjusted for inflation, thus requiring reviews for smaller projects that formerly would have fallen below the threshold. These interviewees specifically recommended that Australia consider inflation adjustments.

One commercial firm that we connected with focused on a red-teaming approach to contestability. Our interviewee felt that reliance on industry partners and leaders has greatly enhanced the firm's risk-identification and -mitigation process. Another commercial firm uses an enterprise-level scrutiny function to consider both the strategic and risk considerations for large future investments.

Institutionalisation

Most scrutiny organisations came about as a result of policy decisions, meaning that they could be modified, replaced, or eliminated without a legislative action. The strongest institutionalisation is the U.S. DoD's CAPE, which is enshrined in law. Policy-based institutionalisation likely reflects the fact that scrutiny *organisations* as permanent fixtures are relatively new developments, in most cases. The scrutiny *function* has existed for much longer but has been conducted on more of an as-needed or ad hoc basis. That said, none of the contestability functions with roots in policy indicated a concern that the survival of the function was at risk.

Structure

The Netherlands has a process-focused review that pulls in a set of defined stakeholders who vary depending on the nature of the decision, and who make up what amounts to a virtual team that undertakes the review. Sweden's review function engaged in oversight rather than formal scrutiny. The stand-alone offices also varied from ones that conducted reviews in-house to ones that facilitated a review function, managing the coordination of the effort and pulling in the right analysts. That said, even the organisations focused on facilitating benefit from some level of expertise on the topics at hand to ensure that the work is being performed properly. In each of the above structures, external expertise can be used on an ad hoc basis to help supplement some of the scrutiny functions (e.g., a specific study on a weapon system's affordability or alternatives).

From the expert discussions, we found that it is not uncommon to have multiple contestability functions at various levels and multiple types of contestability functions within each larger organisation (e.g., the UK MoD or the U.S. DoD). There are also instances where contestability functions exist at lower levels of the organisations for smaller investment decisions that were not as "formal" and independent as some of the functions described above. There also are contestability organisations completely independent of the larger organisations that are required to come in and perform an audit, such as in the case of the U.S. GAO or the UK NAO.

A few interviewees raised issues relating to the structure of the rest of their defence organisations. One factor was the extent of jointness in their military services. Our interviewees generally praised having strong individual service cultures, regardless of whether they were joint. However, most often procurement was done by a central procurement organisation rather than by individual services. Trades among services capabilities were thought to be easier in those cases. This is outside the control of the contestability function, but jointness will shape the number and depth of the connections between the function and those whose decisions are being reviewed. Furthermore, jointness seems to enable trades among capabilities. It is much easier to review the quality of specific decisions and recommend that programs proceed or get reduced or cancelled than it is to perform portfolio reviews and make trades among weapon systems, especially when these are in different services.

In most of the cases, the contestability functions do not have direct decisionmaking power but report directly to a higher authority that does.

Types of Engagement

For contestability, the types of engagement varied, across the organisations we spoke with, from annual, to periodic, to ad hoc as needed (e.g., a private-sector red-teaming exercise for a specific, high-risk investment decision). Our UK interviewee described interactions conducted as needed to support specific committee meetings, with a focus on improving the outcomes of programs. Other organisations offered reviews at specific acquisition gates or milestones (as in the Netherlands, New Zealand, and Canada)—sometimes (as in the Netherlands) during the initial requirements determination phase, as well as at the acquisition gates. Another option, which we saw in the private sector, was a yearly review of projects as part of budget development and investment portfolio allocation, although this seemed to be more related to oversight than scrutiny. Thus, the variance in engagement seems to be linked to what the organisation is set up to do. If it is focused on challenging decisions as a separate process, then engagements will happen after the recommendations are made. If an organisation is attempting to improve the decision while it is in process, more-frequent engagements may be required.

Funding

None of the interviewees at organisations with formal, stand-alone contestability structures expressed concern about funding availability or the source of the budget for the function; although, in some cases, workloads could have been eased with the addition of more staff. The size of the organisation was typically related to the number of decisions that needed to be reviewed or the need to maintain a lean, additional corporate-level function in the private sector. That said, in spite of the size of CAPE (160 staff members), the significant workload often requires long hours. One country with no independent scrutiny function (Sweden) has a relatively small MoD, and we speculate that top-line funding limits contribute to the absence of an office and investment portfolio allocations. This may also be the case for Denmark and the Netherlands.

The commercial firm using red-team reviews viewed inputs from the expert staff it engaged as highly valuable and likely to save more than the cost. This outcome stands in contrast to defence organisations, which do not have metrics, such as return on investment, to justify the costs of contestability.

No defence organisation—including the largest (CAPE)—had adequate resources to execute all of its desired analyses in-house. Generally, they pulled from across their defence departments and often hired specialised contractors to provide supporting analyses. They commonly had budget control to make decisions about where to spend their analytic resources but did not have unlimited funds.

Outputs and Recipients

Typically, the outputs of contestability reviews were recommendations or contributions to decisions made by others, usually senior leadership, rather than the final decisions themselves. Some of the output was for broader audiences—as is the case for GAO or NAO, which have audit functions that post reports on external websites—but most appeared to be for internal government purposes. Budget authority is generally retained by elected officials, who thus maintain the final decisionmaking authority, but there are differences in how involved they get in specific decisionmaking.

CAPE makes recommendations to the U.S. Secretary of Defence and the Deputy Secretary of Defence, who have decision authority. In the Netherlands, final acquisition decisions are made by the Minister of Defence, who also weighs in more directly when the review process does not yield an agreed-upon decision. On the other hand, in New Zealand, the contestability review teams have the authority to cancel the program. In the private sector, the contestability functions inform CEOs, CFOs, and boards of directors, who have the final decision-making power.

Ultimately, decisions can be affected by issues, such as political questions, that may be beyond the purview of the contestability function and even beyond the control of the leaders of the defence organisations that the function informs. In other words, the final decisions of senior defence officials can sometimes be overruled by parliaments or congresses. This contributes to the challenge of measuring the effect of the contestability function.

Standards

Our interviewees indicated the necessity of high-quality reviews, although they did not offer many specifics. The defence organisations tried to use what they understood as best practices for review, but our interviewees did not point to specific standards, models, or methodologies.

Standards relate to another finding. Most defence contestability functions did not generate independent estimates for their reviews. Instead, they use as their starting point data provided by the program or program developer; they then assess the quality of those data. In the United States, CAPE has the scope, resources (in the form of a standing cost-assessment division), and legal mandate to generate independent estimates. CAPE collects data directly from prime contractors and maintains a database that it uses throughout the review function. Other organisations also use a range of commercial tools to assist in performing their analyses.

The collection, storage, and use of data for these functions are a challenge across the board for both public and private organisations. While the functions are collecting large amounts of information for each analysis, there do not appear to be consistent practices or processes in place to capture and store all of the information related to each decision. In most cases, it was not clear that metrics were maintained and reviewed. Some functions try to collect and analyse their own data, while others collect the information from outside (e.g., program offices) and conduct the analysis based on that information. Collecting data separately maintains better independence but can be expensive and difficult.

An even greater challenge is that of retroactively reviewing the decision and the review to assess whether the decisions were correct and the reviews made a difference. One finding is that once the decisions are made, the contesting functions identified in this analysis tend to not go back and track to what extent their recommendations were accepted. This is particularly true for the government functions. It is not entirely clear whether this is because of resource constraints or lack of procedures and processes in place to follow up.

Staffing

In all of the cases we reviewed, formal and informal contestability functions were staffed with experienced senior people with a range of expertise, and interviewees mentioned the importance of having talent with the right analytical experience. Some of the specific types of expertise mentioned were strong analytic backgrounds from academia, previous program managers, military experts, cost analysts, lawyers, members of industry, technical experts, and lifelong civil servants. In defence organisations, these senior staff members were most often civilians; if they were military, they were in the senior ranks. One interviewee expressed regret that the organisation was too small to include developmental staff members, while others celebrated their ability to benefit from pulling in very senior experts who had gained experience in other parts of the organisation.

Smaller organisations admitted that they could not maintain the full range of capability on staff, and even the largest one (CAPE), which maintains expertise across the mission set, also uses outside experts for specific technical questions. External advice and skills are frequently sought to fill gaps that exist in the internal workforce. The use of external expertise in supplementing the contestability function is a theme across nearly all of the interviews, across different structures, and in both private and public sectors. Approaches to bringing in outside expertise varied. Some were able to hire outside support to get the expertise they needed, while others turned to experts inside the government. CAPE mentioned a few preferred providers, and the U.S. federally funded research and development centre (FFRDC) structure enables CAPE to engage support quickly, without a lengthy procurement process. The Netherlands indicated that a benefit of its virtual review structure is that every review can be staffed with relevant experts. New Zealand brings in experts from its allies (including Australia and the UK) to support its reviews. The private companies in this analysis also mentioned bringing in external industry experts as needed.

Staff members in contestability functions are usually very experienced, have worked many years in a field, and have deep understanding of the topics at hand. Advanced degrees are common. Staff respect and support the culture associated with a contestability function (e.g., unbiased analysis). The workload was heavy at many organisations, with staff working long hours.

Not surprisingly, projects are staffed most often with analysts or outsiders who have expertise in that particular area. Internal portfolio management within the contestability function is important to ensure that resources are appropriately matched to questions.

Leaders of the contestability functions were either political appointees or civil servants, depending on how the function was structured.

Incentives

No specific monetary incentives were cited as driving behaviours, nor were any incentives specifically identified that could do so. Rather, we heard that making contributions to the organisations' missions was independently rewarding to the staff members. The shipbuilder we interviewed, and whose organisation used red teams, indicated that the experts it brought in as short-term consultants enjoyed making contributions to decisionmaking and being engaged with the challenging problems; financial incentives were less of a motivating factor for red-team participants.

Review Culture

All discussions of the contestability function that touched on culture stressed the importance of being independent, impartial, and driven by evidence and data. However, where there were formal reviews, there was a variation in the type of review culture described. Our CAPE interviewee stressed the staff members' independence and focus on objective analysis; there was a recognition that analyses could be highly contentious. The Netherlands interviewee described the challenge function reviews as a "dancing table," where the various parties tried to pull the decision into their corner of the room. Sometimes it was highly contentious, while other times friendlier. The UK interviewee for the scrutiny function described a much more collaborative process where the organisation offered independent reviews of work in progress to improve the outcome, rather than end-game reviews. Culture, similar to engagement patterns, seems to match mission, but the topic of the review sometimes shaped how contentious the discussions were. Some topics are inherently more controversial, and those reviews will be difficult.

One private-sector firm suggested that honest and blunt reviews were immensely valuable and that the typical red-team review yielded a few useful points that shaped strategy and decisionmaking, while another mentioned that the organisation understands the value of the contestability function and supports it.

Throughout the interviews, it appeared as if contestability functions located internally to the defence organisations are more likely to be supported as necessary and for the good of the organisation, while external functions (e.g., GAO and NAO)[2] can be looked at as more of a hindrance to getting work done.

---

[2]   OMB was not one of our case studies, but it also performs certain contestability functions directly for the U.S. President. It too is sometimes viewed by other agencies as an external hindrance.

Metrics

We did not uncover any standard metrics for public organisations. Rather, the focus was typically on improving the quality and robustness of decisionmaking and, for the government agencies, ensuring that public funds were well spent. Interviewees and the literature also highlighted the perceived value of being able to better defend decisions to senior officials and the public. The interviewees in government contestability organisations effectively communicated the gravity of their mission. We note that the literature recognises the challenge of measuring outcomes in public institutions, but in cases where advice was being provided directly to senior leadership for decisions, the metric typically measured whether senior leadership was given sufficient or accurate information to make an informed decision. Our findings align with this.

Commercial companies use common financial metrics, including rate of return on investment, and generally are able to take a nearer-term focus.

Risk and Risk Management

Contestability functions are frequently used because the projects or programs are high risk, or because projects, programs, or senior leadership need assistance in understanding the risk—thus hoping to avoid failures that could affect the whole portfolio. However, risk is also usually thought about by programs during the acquisition life cycle. Both public organisations and commercial companies monitor and take account of risk in their decisionmaking, with financial risk being the focus of the latter group.

Buying commercial off-the-shelf systems is seen as a way to reduce or manage risk, but cost growth and unexpected costs remain possibilities even when not developing new systems.

Defence organisations identified a variety of different programmatic risks, including those relating to financial, technical, personnel, and infrastructure, and have different means of exposing them. Germany requires these to be deliberately listed with mitigation plans. Sweden suggested that buying commercial off-the-shelf technology offers a way to manage risk; however, the consensus was it was impossible to eliminate all risk.

Summary Considerations

Looking across our case studies, we found no clear "right" answer or best organisational practice in contestability. Each approach was unique, and we were not able to group the organisations we talked to into a discrete set of archetypal models. This may have been a result of small sample size—if more organisations were included, then clearer patterns may have been revealed. Or it may be that even a larger sample would reveal that funding availability, local political context, history, organisation culture, or some other factor would make each organisation's approach unique to the local ecosystem, no matter the size of the sample. That said, it is possible to summarize our broad findings on the various aspects of contestability, which are presented in Table 4.1.

Table 4.1
Findings on Aspects of Contestability

Aspect	Insight
Function	• Function varies by organisation • Function focuses on militarily significant investments • No organisation reviews every single investment decision • Industry tends toward a red-team approach
Institutionalisation	• Most organisations are formed from policy decisions • Scrutiny functions have predated the existence of formal scrutiny organisations • Functions not seen as at risk
Structure	• Structure varies substantially in size • Biggest discriminator is stand alone or internal to another process • Stand alone varies: some organisations undertake reviews and some only facilitate them • Many contestability functions are common at different levels • Jointness[a] facilitates (or forces) trade-offs
Types of engagement	• Type of engagement varies as a function of what organisation was created to do: challenge decisions or improve decisionmaking processes • Organisations do not review all decisions

Table 4.1—Continued

Aspect	Insight
Funding	• Funding is not a common concern • Commercial red teams save more than they cost • No one has enough staff to do all analyses desired
Outputs	• Outputs include recommendations to others • Approve or cancel program investments (in some cases) • Outputs often can be overturned
Standards	• High-quality standards • Most defence contestability functions do not generate independent estimates for their reviews; they start with program/project staff input • Collection and storage of data is a challenge • Retroactive reviews are difficult (how do you know whether it was a good decision?)
Staffing	• Personnel are highly experienced • Staff are mostly civilian, but, if military, very senior • Use of external expertise is common • Most organisations are often staffed with analysts or outsiders with expertise in specific areas
Incentives	• Contributions to organisation mission are commonly cited • Financial incentives are never cited (even for red-team participants)
Culture	• Independence valued
Metrics	• There are no standard metrics • Typical focus is to improve quality and robustness of decisionmaking and ensure that public funds are well spent • Metrics provide better defence of decisions
Risk	• This aspect is frequently used because senior leaders need to understand risks • Both public and private organisations take account of risk in making decisions, with financial risk the focus of commercial organisations

[a] The degree of integration among different military services.

From these findings, we derive some important aspects of any contestability function that should be part of the initial design:

• Ensure that participants have a sense of independence.

- Ensure that participants are able to offer inputs without fear of retribution.
- Ensure that there are adequate resources so that all decisions that reach whatever threshold is used to require review can be reviewed; there should be no chance for a biased selection of what to analyse.

Additionally, based on this research, we see several key decision points for Australia to consider when choosing particular aspects or characteristics of the ADoD contestability function. The starting assumption is that there will be some sort of function. Once that is certain, some reasonable questions follow:

- Should there be a standing contestability organisation with the responsibility of performing the reviews, or should there be a review structure with a small footprint that can pull teams together on an as-needed basis? If a standing organisation, should it be stand-alone or incorporated into an existing function or structure? And how does the choice affect the independence of the contestability function?
- Will the organisation have a permanent staff large enough to do all the work in-house or will it sometimes need to engage outside staff? Will the staff consist of government officials or outside contractors, or will it depend on what expertise is required? Or will the organisation have a small staff that facilitates reviews and designs each review separately? How can it be ensured that this staff has the resources necessary to accomplish the mission?
- To whom does the contestability organisation report and what are the outputs? What is the nature of the interrelationships among the oversight and decision authorities in the ADoD and national-level oversight (e.g., Parliament) and audit functions? To what extent does the contestability organisation make final decisions (e.g., to cancel programs) as opposed to advising higher-level decision authorities?

- Will the decisions be contested at the decision gates? Or will the organisation work to improve ongoing decisionmaking? What thresholds will trigger a review?
- Will the contestability organisation be given unfettered access to all data, information, and reports collected during the initial decisionmaking process, or will it have its own independent data sources—or both? And will the organisation under review be required to respond explicitly to the contestability organisation's findings? Will it review outcomes over the longer term to see the impact of decisions, as a source of lessons learned for the function?
- Will the function play a role in maintaining and managing data?

Organisational Change

Depending on how it is structured, the adoption of a contestability function in the ADoD could have a fairly profound effect. If the outcome of the contestability review means that programs are approved—or cancelled—or if force structures are changed, or even if it merely requires further analysis be done, then this represents a new source of authority. How can Australia ensure that this new effort is successful? As one of our interviewees put it, "Changing wiring diagrams is easy, changing culture and process is harder." The addition of a new contestability function may lead to a new block on the organisational chart and some new steps in the decisionmaking process, but that is not enough to ensure that the desired outcomes are met.

The addition of a contestability function represents organisational change, and consequently can be managed formally as one. To increase the chances of ensuring that the new contestability function helps improve decisionmaking and resource allocation, the ADoD can adapt and adopt principles of successful change as it proceeds. This analysis focused more on alternatives for structuring different aspects of the contestability function, rather than how to institutionalise it. However, this final section provides some thoughts on organisational change.

There is an extensive academic literature on change that offers insight into how to effectively manage change. Back in 1947, Kurt Lewin published what we might now think of as an obvious truism, that there are three main stages for change: preparing, enacting, and consolidating (or as Lewin put it, unfreezing the current situation, moving or changing, and refreezing).[3] John Kotter has offered perhaps the best-known structure for the steps of change.[4] These are:

1. creating urgency
2. creating a coalition
3. developing vision and strategy
4. communicating the vision
5. empowering employees
6. generating short-term wins
7. consolidating gains
8. anchoring the change in culture.

The final model we offer is by Fernandez and Rainey,[5] who focus on the public sector and again offer eight steps:

1. ensure the need
2. provide a plan
3. build internal support for change and overcome resistance
4. ensure top-management support and commitment
5. build external support
6. provide resources
7. institutionalise change
8. pursue comprehensive change.

---

[3] Kurt Lewin, *Frontiers in Group Dynamics*, The Bobbs-Merrill Reprint Series in the Social Sciences, S170, n.p., 1947.

[4] John P. Kotter, *Leading Change*, Boston, Mass.: Harvard Business School Press, 1996.

[5] Sergio Fernandez and Hal G. Rainey, "Managing Successful Organisational Change in the Public Sector," *Public Administration Review*, Vol. 66, No. 2, 2006.

These and other models offer insights into necessary aspects of organisational transformations, but they all must be adapted to the context in which the change is occurring.

Several of these steps are in place already. Australia has plans to go ahead with a contestability function. The need for change is laid out in *First Principles Review: Creating One Defence*. Structuring the change (the *what*) will not necessarily be a simple task—indeed, this report is specifically aimed at providing support for that process and shows that there are a number of reasonable alternative structures and subprocesses to select from—but once these decisions are made, the focus needs to be on the *how*. The existence of *First Principles Review* is a signal that the need is there, and it is a signal of senior leadership support.

Senior-management support—by which we mean support from the top of the ADoD, not internal to the new contestability function—is critical to any change. Indeed, without it, much of the other fundamentals of successful change will be weaker. Top-management backing will be required in setting up the function within the organisation and process structure, including the new structure and the review requirements, and is necessary to ensure support from the rest of the organisation. Types of support required include adequate resourcing in terms of having the right number and mix of employees, support when those employees pull in expertise from across the ADoD, and adequate resources to hire external support if needed. Also, management support must be consistent, with the goal of imbuing contestability in the culture and having all relevant organisations either agree that it is a value added step or at least accede to the necessity of doing it. This means that once the framework for review and the decisionmaking threshold are determined, the organisation must support these on an ongoing basis. Allowing exceptions for programs to avoid review will fundamentally weaken the function and create the danger that stakeholders will focus on arguing for exceptions rather than ensuring that their analyses are solid.

Another commission for senior leadership is to help ensure that, to the extent possible, the recommendations from the contestability function have real influence. If the recommendations are routinely ignored, then the effect of the function will be limited. There may always be political reasons why a decision that is viewed by the contestability organisation as most cost-effective (for example) is not the preferred option of senior leadership, but the reasons should be clearly understood and should not strictly be the preference of the originator of the decision that was reviewed. Culture is an important aspect of this, an internal culture of independence and objectivity, as well as the overarching culture of the organisation, to truly value what independent reviews can offer.

The plan for change should include instituting all the various features as ultimately decided by the ADoD, but there will remain open questions to address. One consideration is the different challenge of instituting contestability for brand-new programs, compared with ones that are under way. Should all in-process programs be reviewed, and how should challenges of existing programs be dealt with? Some advance thought to questions like these would be useful. There are many details to be thought through, and one approach to identifying them would be to hold an experimental review geared toward looking for points where further definition is necessary.

The final aspect of any change is that it can be monitored and reviewed on an ongoing basis. Nations with long-standing contestability functions in their defence organisations have typically evolved them over time. Whatever initial contestability approach Australia decides on can be examined after some time has passed and refined to take advantage of lessons learned from experiencing these reviews.

Summary Principles

We close by offering some summary principles for the incorporation of a successful contestability function, listed in the text box.

There are many choices in contestability approach and organisational design. Specific details of strong contestability functions do vary, but these principles will enhance the success of the contestability function over time and will help create a strong and effective system of checks and balances to ensure the best allocation of public resources in the defence of the nation.

Summary Principles for Incorporating Contestability

Senior-leadership and line-manager support for new contestability function

Clear mission and understanding of what resource decisions the contestability function will review and where it is injected into the process

Clear understanding of the outputs and goals of the contestability reviews

Ongoing leadership support for contestability recommendations as quality inputs are to be taken very seriously (even if final decisions are different)

Whomever the contestability function reports to has real decisionmaking power

Adequate resources for the contestability function

Adequate staff of senior, experienced experts for the contestability function

Timely access to the right data

Development of an independent review culture, without fear of retribution

Storage of analysis and decisions over time to create a body of knowledge, which will help to increase the long-term success of the function

Contestability Protocol

This appendix reproduces the protocol that we used in our discussions with experts.

Thank you for agreeing to participate in this data-collection effort. As we described in our outreach, we work at the RAND Corporation, which is a nonprofit, nonpartisan research organisation. Our core values are quality and objectivity, and our research is disseminated as widely as possible to benefit the public good. Although best known for the independent analysis we provide the U.S. Department of Defense, our research activities span a much broader spectrum of topics and clients, including nondefence analysis for many sponsors and defence research for allied nations.

Recently, RAND became involved in the implementation of recommendations from Australia's *First Principles Review: Creating One Defence*. As a part of that implementation, the ADoD has asked RAND to investigate global best practices in *contestability*, which refers to an independent review function for decisionmaking, with the goal of providing the optimum portfolio of military capabilities through the most efficient and effective use of public funds. The function is intended to provide assurances to the senior leadership and to the government that the ADoD's capability needs and requirements are not only in alignment with strategy and resources but can also be delivered effectively. The ADoD has asked RAND to reach out to organisations to learn about best practices and models that have underpinned successful contestability efforts in both the public and private sectors.

Before we begin this discussion, I would like to confirm that you are participating in this on a voluntary basis. (Yes, no. If no, end the conversation.)

A bit about the ground rules: If you prefer, you will not be cited by name. For private-sector companies: Would you prefer that we do not include the name of your company? We can describe you as "a firm in *X* industry."

1. Does your company or organisation have the type of organisation(s) that do these types of reviews? What is the history of the contestability function in your organisation?
   a. When was this function established?
   b. Was there an event that led to the formation of this function?
   c. Is there a vision or mission statement, or a formal charter or guiding principles that outline some of these functions? (If yes, please provide.)
2. Before we talk in more detail about the contestability reviews, let's talk first about the kind of decisions that are subject to these reviews. We will refer to these as *resource decisions/recommendations*. Can you please describe these? (If appropriate: How do these connect to the desired outcome/mission of your organisation?) (This could include big capital investments, decisions about force structure allocation, and decisions about whether or not to invest in a major weapon system.)
   a. What are the steps in the initial resource decision/recommendation making process? Do you have a document outlining this?
   b. How is the "requirement" for the new capability determined? Who makes the decision about this? (Private-sector CEO? Board of directors? Others?)
   c. Who selects the alternatives under consideration and how are the alternatives selected for assessment?
   d. What decisionmaking processes are followed to make resource decisions/recommendations? What are the important metrics and who chooses the metrics?

    e. Who makes the final resource decisions/recommendations? (Same as the requirement generator, or another entity?)

3. After resource decisions/recommendations are made, how are they subject to "contestability" (or scrutiny/review, etc.) (We will refer to this process as being *contested* to clarify the difference from the initial resource decisions/recommendations.)

    a. Which resource decisions/recommendations are contested— is it everything? Some subset based on a cost or other metric or a subjective assessment? Who selects which are contested?
- How many decisions are contested every year? (What is the workload of the function?)

    b. Who or what organisation(s) does the actual contestability review? How is this done analytically?

    c. How is the contestability-reviewing organisation structured?
- Are the resource decisions/recommendations in the review contested in the same department as they were made initially or in a different department?
- Who manages the function? How is the leadership determined (e.g., elected, appointed, career, political)? What is their typical background or expertise? Is there a term limit (i.e., rotating function) or is it a permanent post?
- What is the reporting structure of the original resource decision/recommending organisation and the contestability organisation?
  - Do they report to the same group? Who has final oversight over these teams?
  - Who specifically does the contestability function report to? At what level—is it to a line-management group or to the top of the organisation? Are they within the same group or organisationally separated? (If possible, please provide an organisational structure chart identifying these positions.)
  - Is there a reclama process for disagreements?
  - How is the independence of the contestability review team maintained?

    ◦ What is the authority of the contestability review function? Can they force a program to stop or do they have to go to a third party?

4. Using external sources of analysis—what kind of external expertise is used, if any?
   a. Which decisions are subject to further external review?
   b. What types of organisations conduct the external reviews?
   c. How does this analysis support decisionmaking?
   d. How does internal staff interact with external agencies, groups, organisations?
5. Contestability review management
   a. How are projects internally categorised and assigned? (Dollar or size or expertise?)
   b. Is it a continuing function? Or episodic?
   c. What data are used to contest the resource decisions/recommendations?
      ◦ How is this developed, maintained, validated?
   d. Are there single points of review? Or multiple points along a resource decisions/recommendation project life cycle? (In the United States, defence programs are reviewed multiple times at milestones. In Australia, they are reviewed at gates, which are similar to U.S. milestones.)
      ◦ How is continuity of review or expertise maintained? Do the same people or organisations review resource decisions/recommendations through their life cycle?
      ◦ How are data handled, managed, and stored throughout the contestability review life cycle if there are multiple decision points?
   e. What incentives are used for the contestability review function to promote its mission?
6. How is risk handled?
   a. What are the types of risk that are assessed (e.g., financial, technical)?
   b. What are the metrics for assessing risk?

   c. Is risk reviewed by the same people who do financial reviews or is there a separate organisation or function (e.g., for technical risk)?

     ◦ How is residual risk managed?

      – How does the organisation balance cost and capability trade-offs in a risk framework?

7. How is the contestability function funded and staffed? Where do these funds come from? (Possibilities: defence budget, overhead, etc.)

   a. How is the budget determined? Is it limited? Are there cases where you can't review decisions because of insufficient funding?

   b. What is the annual budget for this function?

   c. How is the organisation or function staffed?

     ◦ How many people are employed by the function?

     ◦ How are staff identified and recruited to join the function?

     ◦ What is their typical background or expertise?

     ◦ How do contestability staff interact with the rest of the organisation?

   d. What is the culture of the organisation? How is that maintained and strengthened?

8. Contestability outputs: How are the results reported and to whom? What are the typical outputs of this function?

   a. Are the efforts integral to the decision process, or is it a special effort to shed light on decisions but not directly influence events?

   b. Do you have an example of how the contestability function supported a contentious decision? Or how it rejected a decision?

   c. How is success of the analysis reviewed and measured?

   d. (If not answered previously, ask about a reclaim process.)

9. What have we missed? Are there any other points about contestability functions that weren't touched on here?

Examples of Contestability Functions from the Literature

Contestability functions offer a number of different formal ways to undertake independent reviews of significant recommendations on large-scale movement of resources. Each example highlights different functions. In this appendix, we present a set of examples of contestability functions we found in the literature by way of illustrating the variety of forms and information available on these functions. In some cases, the examples provide the key recipe for success in the *specific circumstances* in which they are applied or some advice on what to consider or how to go about conducting oversight.

Centre for Public Scrutiny: *The Good Scrutiny Guide*

The Centre for Public Scrutiny in the United Kingdom "outlines four principles of effective scrutiny which together build towards an accountability cycle to improve public services."[1] The London Borough of Merton used these principles to provide guidance on effective public scrutiny. Some of the pertinent information translates over to a scrutiny function when making resource decisions:

> The "critical friend" role of scrutiny is founded on a mutual respect relationship between scrutiny and the executive.

---

[1] Centre for Public Scrutiny, *The Good Scrutiny Guide: Pocketbook for Scrutineers*, London, undated.

For scrutiny to be effective its status must be recognised as being on a par with that of the executive ("parity of esteem").

There needs to be:

- Clear rules of engagement.
- Coordinated workload planning with clear link to corporate processes, dovetailing the work of scrutiny with policy development and decision-making cycles to maximise influence.
- Reporting and monitoring mechanisms

Need to acknowledge that scrutiny is an ongoing process: stated outcomes should be monitored and outcomes should be reassessed where expectations have not been met.

Behaviours for scrutiny to make a difference:

- Confidence
- Credibility
- Legitimacy
- Command attention
- Influence
- Persuade
- Challenge
- Non-aggressive

Supported by:

- A high calibre of practitioners who use objective questioning, are clear about what they want to know and plan their questioning around these goals
- Inclusive and focused chairing.[2]

---

[2] Overview and Scrutiny Commission, *Overview and Scrutiny at the London Borough of Merton*, London: London Borough of Merton, March 2006, p. 9.

The Institute of Internal Auditors: *Supplemental Guidance: The Role of Auditing in Public Sector Governance*

The Institute of Internal Auditors

> presents information on the importance of the public sector audit activity to effective governance and defines the key elements needed to maximize the value the audit activity provides to all levels of the public sector. The guidance is intended to point to the roles of audit (without differentiating between external and internal), methods by which those roles can be fulfilled, and the essential ingredients necessary to support an effective audit function. As such, it may not be fully applicable in every jurisdiction, particularly where public sector audit roles and responsibilities are specifically defined by governing institutes or legal mandates to exclude certain functions or assign them to other entities.[3]

The institute concluded that all public-sector audit activities require these characteristics:
- Organisational independence
- A formal mandate
- Unrestricted access
- Sufficient funding
- Competent leadership
- Objective staff
- Competent staff
- Stakeholder support
- Professional audit standards.[4]

---

[3]   Institute of Internal Auditors, 2012, p. 3.

[4]   Institute of Internal Auditors, 2012, p. 8.

Project Management Journal: "Governance Frameworks for Public Project Development and Estimation"

The authors of an article in the *Project Management Journal* did "a systematic comparison of the governance frameworks developed within the United Kingdom and Norway for large public projects, their history, and as they were in mid-2007."[5] In particular, the authors looked at the Norway Ministry of Finance, the UK MoD, and the UK Office of Government Commerce frameworks. Some common best practices in governance principles were mentioned:

- Transparency, openness for scrutiny, maximum openness about basis for decisions
- Learning, willingness to change
- Setting common, high professional standards
- External control, independency
- Political anchoring of framework on high level
- QA [quality assurance]/Gateway review is non-political
- Look for big, important trends, not the minor details.[6]

General Electric Capital: *Due Diligence: Main Steps and Success Factors, Overview*

General Electric Capital published the main steps and success factors in a guide entitled *Due Diligence*.[7] The background for this function is the following:

> For acquiring companies, due diligence is a critical process that cannot be overlooked. Due diligence not only nets the hard data you need to assess potential financial, legal, and regulatory exposures, but also gives insights into the target company's structure,

---

[5] Ole Jonny Klakegg, Terry Williams, Ole Morten Magnussen, and Helene Glasspool, "Governance Frameworks for Public Project Development and Estimation," *Project Management Journal*, Special Issue: Special PMI Research Conference, Vol. 39, No. S1, 2008.

[6] Klakegg et al., 2008.

[7] General Electric Capital Corporation, *Due Diligence: Main Steps and Success Factors, Overview*, Norwalk, Conn., 2012.

operations, culture, human resources, supplier and customer relationships, competitive positioning, and future outlook. Done right, due diligence is a way to spot potential deal-killers/shapers and provide assurances that the acquisition is the right decision at the right price. Done fully, due diligence can also give management a deep, holistic view of the target company that can later inform integration of the target's people and business.

Some of the key inputs for this due diligence function include
- form and begin prepping your due diligence team
- bring in outside expertise as necessary
- get the integration manager involved early
- create due diligence checklists
- prepare your data requests
- negotiate and sign a confidentiality agreement
- establish and index a physical or online data room for confidential documents
- prepare a communication plan.[8]

U.S. Securities and Exchange Commission: "Investment Advisor Due Diligence Processes for Selecting Alternative Investments and Their Respective Managers"

The U.S. Securities and Exchange Commission (SEC) published a risk alert to draw awareness to the following:

For at least the past six years, staff in the Office of Compliance Inspections and Examinations (the "staff" and "OCIE" respectively) have observed and outside studies have indicated that investment advisers, including pension consultants, are increasingly recommending alternative investments to their clients. . . . The due diligence process can be more challenging for alternative investments due to the characteristics of private offerings, including the complexity of certain alternative investment strategies. The staff examined the due diligence and related investment advisory

---

8   General Electric Capital Corporation, 2012.

processes of certain advisers to pension plans and funds of private funds in order to evaluate how these advisers: (i) performed their due diligence; (ii) identified, disclosed, and mitigated conflicts of interest (e.g., benefits to the adviser or its employees for allocations made to private funds); and (iii) utilized experienced investment teams when evaluating complex investment strategies and fund structures.[9]

The SEC then listed practices

employed by some advisers that may provide greater transparency and that independently support the information provided by underlying managers[, which] include: (i) the use of separate accounts to gain full transparency and control; (ii) the use of transparency reports issued by independent fund administrators and risk aggregators; (iii) the verification of relationships with critical service providers; (iv) the confirmation of existence of assets; (v) routinely conducting onsite reviews; (vi) the increased emphasis on operational due diligence; and (vii) having independent providers conduct comprehensive background checks.[10]

Ernst & Young: *Third-Party Due Diligence: Key Components of an Effective, Risk-Based Compliance Program*

Ernst & Young produced a document on due diligence:

The economic crisis, vigorous governmental enforcement activity and the increased focus on enterprise risk are causing global corporations and their audit committees to take a closer look at how they manage and conduct their due diligence around vendor,

---

[9] U.S. Securities and Exchange Commission, Office of Compliance Inspections and Examinations, "Investment Adviser Due Diligence Processes for Selecting Alternative Investments and Their Respective Managers," *National Exam Program Risk Alert*, Vol. 4, No. 1, January 28, 2014.

[10] U.S. Securities and Exchange Commission, Office of Compliance Inspections and Examinations, 2014.

distributor, joint venture and customer organisations—defined broadly as third parties. Those with existing due diligence programs are finding they have not kept up with the increased global risks of third-party vendors—particularly in the areas of anti-bribery and corruption—leaving many companies to wonder what constitutes a reasonable due diligence program and how much research and documentation are enough.[11]

Ernst & Young created a due diligence methodology "to provide adequate risk-based categorisation, appropriate levels of data analysis, ongoing monitoring and effective communication" for vendors, distributors, joint ventures, and customer organisations.[12] The company provided the following key questions to ask:

Consistency. Is the process followed consistently? Can you audit or tie back vendor request forms to each vendor in the vendor master? Is there training around the process? Is it globally deployed? Is the process repeatable—i.e., would you arrive at the same conclusion if you were to run a selection of new vendor setup forms through the same process? Are the rules and contract language around FCPA and anti-corruption consistent from country to country?

Management oversight. When was the last global training program on anticorruption, due diligence or compliance? When did you last update your new vendor setup form or procedures? Does your company use software tools for case management to manage and document the vendor setup process? What database and due diligence steps does accounts payable take to categorise new vendor submissions received from the requestor? Is the right person making the decision? Once accepted, is it rechecked annually or on an ongoing basis? During the escalation process, who is responsible for making the tough calls? How robust is the vendor "vetting report"? Does it incorporate public database

---

[11] Ernst & Young, Australia, *Third-Party Due Diligence: Key Components of an Effective, Risk-Based Compliance Program*, Adelaide, 2012.

[12] Ernst & Young, Australia, 2012.

checks, include the officers of a company and search for "politically exposed persons," adverse media, country-specific sanctions and more? Who is made aware of a new vendor once approved—is it communicated to the corporate office and centrally managed, or is it handled and decided by the local office?

Objectivity. Given so many decision-makers at the country or subsidiary level, can the current process stand up to independence scrutiny from an outside (or DOJ [U.S. Department of Justice]) perspective?

Reasonableness. Is the process reasonable? Does the process generate too much paperwork that may not get reviewed or too little paperwork where rogue third parties or necessary contract terms might be missed? Does the process incorporate leading practices, including the criteria set forth in the U.S. Sentencing Guidelines and OECD [Organisation for Economic Co-operation and Development]?[13]

Congressional Research Service: *Congressional Oversight Manual*

The U.S. Congressional Research Service developed the *Congressional Oversight Manual* in 1978 with the assistance of a number of House of Representatives committee staffers; the manual has been updated as needed since that time.[14] The manual provides very specific information on congressional oversight, including how to conduct various types of oversight. The manual's sections are titled

- "Investigative Oversight"
- "Constitutional Authority to Perform Oversight and Investigative Inquiries"

---

[13] Ernst & Young, Australia, 2012.

[14] Alissa M. Dolan, Elaine Halchin, Todd Garvey, Walter J. Oleszek, and Wendy Ginsberg, *Congressional Oversight Manual*, Washington, D.C.: Congressional Research Service, December 19, 2014.

- "Authority of Congressional Committees"
- "Legal Tools Available for Oversight and Investigations"
- "Enforcement of Congressional Authority"
- "Limitations on Congressional Authority"
- "Frequently Encountered Information Access Issues"
- "Individual Member and Minority Party Authority to Conduct Oversight and Investigations"
- "Selected Oversight Techniques."[15]

---

[15] Dolan et al., 2014.

Summary of Major Currency Conversions

Country	Organisation	Contestability Description	Threshold for Review	Original Currency	Value, Millions		
					AUD	USD	EUR
UK	IAC	Investment level requiring IAC approval	GBP 400 million	GBP[a]	808	613	547
UK	IAC	Investment requiring staff scrutiny	GBP 100 million	GBP	202	153	137
UK	MoD	MoD may approve projects (without Treasury approval) costing less than	GBP 400 million	GBP	808	613	547
Germany	Parliament	Reviews projects with a value greater than	EUR 25 million	EUR[b]	37	28	25
Canada	IRPDA	Third party challenge to projects with value equal to or greater than	CAD 100 million	CAD[c]	105	79	71
Netherlands	MoD	Uses four phase process for review for projects over	EUR 100 million	EUR	148	112	100

Country	Organisation	Contestability Description	Threshold for Review	Original Currency	Value, Millions		
					AUD	USD	EUR
Netherlands	DMO	Use Defense Materiel Process for all projects of value equal to or greater than	EUR 5 million	EUR	7.4	5.6	5
Denmark	DALO	MoD must approve projects costing over	DKK 60 million	DKK[d]	12	9	8

[a] Based on the average daily currency exchange rate between 1 December 2014 and 1 December 2015 of 2.0212 AUD, 1.5339 USD, and 1.3685 EUR to 1 GBP, from Oanda.com.

[b] Based on the average daily currency exchange rate between 1 December 2014 and 1 December 2015 of 1.4770 AUD and 1.1221 USD to 1 EUR, from Oanda.com.

[c] Based on the average daily currency exchange rate between 1 December 2014 and 1 December 2015 of 1.0454 AUD, 0.7949 USD, and 0.7087 EUR to 1 CAD, from Oanda.com.

[d] Based on the average daily currency exchange rate between 1 December 2014 and 1 December 2015 of 0.1981 AUD, 0.1505 USD, and 0.1341 EUR to 1 DKK, from Oanda.com.

Abbreviations

ACAT	Acquisition Category
ADoD	Australian Department of Defence
AUD	Australian dollar
BAAINBw	Bundesamtes für Ausrüstung, Informationstechnik und Nutzung der Bundeswehr (Federal Office of Equipment, Information Technology and In-Service Support)
BKartA	Bundeskartellamt
CAD	Canadian dollar
CAPE	Cost Assessment and Program Evaluation
CEO	chief executive officer
CFO	chief financial officer
COO	chief operating officer
DAC	Defence Audit Committee
DALO	Defence Acquisition and Logistics Organization
DCAPE	Director, Cost Assessment and Program Evaluation

DKK	Danish krone
DMO	Defence Materiel Organisation
EUR	euro
FYDP	Future Years Defense Program
GAO	Government Accountability Office
GBP	British pound
IAC	Investment Approvals Committee
ICG	Independent Cost Group
ICT	information and communications technology
IRPDA	Independent Review Panel for Defence Acquisition
MoD	ministry of defence
NAO	National Audit Office
NZD	New Zealand dollar
OSD	Office of the Secretary of Defense
PA&E	Program Analysis and Evaluation
PPBE	Planning, Programming, Budgeting, and Execution
Rü board	Rüstungsboards
UK MoD	United Kingdom Ministry of Defence
USD	U.S. dollar
U.S. DoD	United States Department of Defense

Bibliography

ADoD—*See* Australian Department of Defence.

Alexander, Ernest R., "Institutional Transformation and Planning: From Institutionalization Theory to Institutional Design," *Planning Theory*, Vol. 4, No. 3, 2005, pp. 209–223.

Amengual, Matthew, "Oversight," in Mark Bevir, ed., *Encyclopedia of Governance*, Thousand Oaks, Calif.: SAGE Publications, 2007, pp. 655–656.

Angwin, Duncan, "Mergers and Acquisitions Across European Borders: National Perspectives on Preacquisition Due Diligence and the Use of Professional Advisers," *Journal of World Business*, Vol. 36, No. 1, 2001, pp. 32–57.

Arena, Mark V., and Lauren A. Mayer, *Identifying Acquisition Framing Assumptions Through Structured Deliberation*, Santa Monica, Calif.: RAND Corporation, TL-153-OSD, 2014. As of October 7, 2015:
http://www.rand.org/pubs/tools/TL153.html

Asare, Thomas, "Internal Auditing in the Public Sector: Promoting Good Governance and Performance Improvement," *International Journal on Governmental Financial Management*, Vol. 9, No. 1, 2009, pp. 15–28.

Australian Department of Defence, *First Principles Review, Creating One Defence*, Canberra, April 2015. As of October 13, 2015:
http://www.defence.gov.au/publications/reviews/firstprinciples/

"Australia Pushing Ahead with Defense Procurement Reform," *Defense Daily International*, Vol. 11, No. 34, November 20, 2009.

BAAINBw—*See* Bundesamtes für Ausrüstung, Informationstechnik und Nutzung der Bundeswehr.

Babbage, Ross, Allan Behm, Michael Clifford, Andrew Davies, Graeme Dobell, Ken Gleiman, Paddy Gourley, Peter Jennings, John O'Callaghan, and Mark Thomson, "Reviews and Contestability: New Directions for Defence," *Australian Policy Online*, May 21, 2015. As of November 11, 2015:
http://apo.org.au/node/54848

Baker, George P., Michael C. Jensen, and Kevin J. Murphy, "Compensation and Incentives: Practice vs. Theory," *The Journal of Finance*, Vol. 43, No. 3, July 1988, pp. 593–616.

Bénabou, Roland, and Jean Tirole, *Incentives and Prosocial Behavior*, Cambridge, Mass.: National Bureau of Economic Research, Working Paper 11535, 2005. As of October 7, 2015:
http://www.princeton.edu/~rbenabou/papers/w11535.pdf

Beutell, Nicholas J., "Organisational Staffing," in Jeffrey H. Greenhaus and Gerard A. Callanan, eds., *Encyclopedia of Career Development*, Thousand Oaks, Calif.: SAGE Publications, 2006, pp. 603–607.

Bevan, Helen, Glenn Robert, Paul Bate, Lynne Maher, and Julie Wells, "Using a Design Approach to Assist Large-Scale Organisational Change: '10 High Impact Changes' to Improve the National Health Service in England," *The Journal of Applied Behavioral Science*, Vol. 43, No. 1, March 2007, pp. 135–152.

Bonner, Sarah E., and Geoffrey B. Sprinkle, "The Effects of Monetary Incentives on Effort and Task Performance: Theories, Evidence, and a Framework for Research," *Accounting, Organisations and Society*, Vol. 27, 2002, pp. 303–345.

Bourgeois, Isabelle, Eleanor Toews, Jane Whynot, and Mary Kay Lamarche, "Measuring Organisational Evaluation Capacity in the Canadian Federal Government," *The Canadian Journal of Program Evaluation*, Vol. 28, No. 2, 2013, pp. 1–19.

Bundesamtes für Ausrüstung, Informationstechnik und Nutzung der Bundeswehr, home page, last updated November 11, 2015. As of November 16, 2015:
http://www.baainbw.de/portal/a/baain/!ut/p/c4/04_
SB8K8xLLM9MSSzPy8xBz9CP3I5EyrpHK9pMTEzDz9gmxHRQDxFVO1/

"The Bundeskartellamt," web page, Bundeskartellamt.de, undated. As of November 16, 2015:
http://www.bundeskartellamt.de/EN/AboutUs/Bundeskartellamt/
bundeskartellamt_node.html

Byler, Daniel, Steve Berman, Vishwa Kolla, and William D. Eggers, *Accountability Quantified: What 26 Years of GAO Reports Can Teach Us About Government Management*, Westlake, Tex.: Deloitte University Press, 2015.

CAPE—See U.S. Office of Cost Assessment and Program Evaluation.

Centre for Public Scrutiny, *The Good Scrutiny Guide: Pocketbook for Scrutineers*, London, undated.

"Chapter Six: Asia," *The Military Balance*, Vol. 115, No. 1, 2015, pp. 207–302.

"Chief of Defence Materiel: Bernard Gray," web page, GOV.UK, undated. As of November 16, 2015:
https://www.gov.uk/government/people/bernard-gray--2

Cuganesan, Suresh, and David M. Lacey, "Developments in Public Sector Performance Measurement: A Project on Producing Return on Investment Metrics for Law Enforcement," *Financial Accountability and Management*, Vol. 27, No. 4, November 2011, pp. 458–479.

"Defence and Armed Forces—Guidance: Ministry of Defence Commercial," web page, GOV.UK, December 12, 2012. As of November 16, 2015: https://www.gov.uk/guidance/ministry-of-defence-commercial

Deis, Donald R., Jr., and Gary A. Giroux, "Determinants of Audit Quality in the Public Sector," *The Accounting Review*, Vol. 67, No. 3, July 1992, pp. 462–479.

Deloitte Touche Tohmatsu, *Contestability, A New Era in Service Delivery Reform*, 2014. As of October 7, 2015: http://www2.deloitte.com/content/dam/Deloitte/au/Documents/public-sector/deloitte-au-ps-contestability-a-new-era-in-service-delivery-reform-091014.pdf

Department of National Defence and the Canadian Armed Forces, "Biographies," web page, undated.

———, *Defence Acquisition Guide 2015*, Ottawa, last updated June 25, 2014. As of November 11, 2015: http://www.forces.gc.ca/en/business-defence-acquisition-guide-2015/index.page

DiTomaso, Nancy, and Julia Eisenberg, "Typology of Organisational Culture," in Eric H. Kessler, ed., *Encyclopedia of Management Theory*, Thousand Oaks, Calif.: SAGE Publications, 2013, pp. 913–915.

Dixit, Avinash, "Incentives and Organisations in the Public Sector: An Interpretative Review," *The Journal of Human Resources*, Vol. 37, No. 4, Autumn 2002, pp. 696–727.

Dolan, Alissa M., Elaine Halchin, Todd Garvey, Walter J. Oleszek, and Wendy Ginsberg, *Congressional Oversight Manual*, Washington, D.C.: Congressional Research Service, December 19, 2014. As of October 7, 2015: https://www.fas.org/sgp/crs/misc/RL30240.pdf

Dunleavy, P., Christopher Gilson, Simon Bastow, and Jane Tinkler, *The National Audit Office, the Public Accounts Committee and the Risk Landscape in UK Public Policy*, London: London School of Economics, Risk & Regulation Advisory Council, 2009.

Ernst & Young, Australia, *Third-Party Due Diligence: Key Components of an Effective, Risk-Based Compliance Program*, Adelaide, 2012. As of October 7, 2015: http://www.ey.com/Publication/vwLUAssets/Third-party_due_diligence/$FILE/Third-Party-Due-Diligence.pdf

European Commission, "European Commission at Work," web page, last updated October 27, 2015. As of November 11, 2015: http://ec.europa.eu/atwork/index_en.htm

Fernandez, Sergio, and Hal G. Rainey, "Managing Successful Organisational Change in the Public Sector," *Public Administration Review*, Vol. 66, No. 2, 2006, pp. 168–176.

Försvarets materielverk, "About FMV," web page, last updated January 1, 2013. As of November 11, 2015:
http://www.fmv.se/en/About-FMV/

Francis, Jere R., "A Framework for Understanding and Researching Audit Quality," *AUDITING: A Journal of Practice and Theory*, Vol. 30, No. 2, May 2011, pp. 125–152.

GAO—*See* U.S. Government Accountability Office.

Gash, Alexander, and John Wanna, "Audit," in Mark Bevir, ed., *Encyclopedia of Governance*, Thousand Oaks, Calif.: SAGE Publications, 2007, pp. 36–40.

General Electric Capital Corporation, *Due Diligence: Main Steps and Success Factors, Overview*, Norwalk, Conn., 2012. As of October 7, 2015:
http://www.gecapital.com/en/pdf/GE_Capital_Overview_Due_Diligence.pdf

Greenberg, Michael D., *Transforming Compliance: Emerging Paradigms for Boards, Management, Compliance Officers, and Government*, Santa Monica, Calif.: RAND Corporation, CF-322-CCEG, 2014. As of August 18, 2015:
http://www.rand.org/pubs/conf_proceedings/CF322.html

Hanberger, Anders, "The Real Functions of Evaluations and Response Systems," *Evaluation*, Vol. 17, No. 4, 2011, pp. 327–349.

Hayward, Matthew L. A., and Warren Boeker, "Power and Conflicts of Interest in Professional Firms: Evidence from Investment Banking," *Administrative Science Quarterly*, Vol. 43, No. 1, March 1998, pp. 1–22.

Heinrich, Carolyn J., "Outcomes-Based Performance Management in the Public Sector: Implications for Government Accountability and Effectiveness," *Public Administration Review*, Vol. 62, No. 6, November/December 2002, pp. 712–725.

Hodge, Billy J., William P. Anthony, and Lawrence M. Gales, *Organisational Theory: A Strategic Approach*, 5th ed., Upper Saddle River, N.J.: Prentice Hall, 1996.

Institute of Internal Auditors, *Supplemental Guidance: The Role of Auditing in Public Sector Governance*, 2nd ed., Altamonte Springs, Fla., January 2012. As of October 7, 2015:
https://na.theiia.org/standards-guidance/Public%20Documents/
Public_Sector_Governance1_1_.pdf

Jarvaise, Jeanne M., Jeffrey A. Drezner, and Daniel M. Norton, *The Defense System Cost Performance Database: Cost Growth Analysis Using Selected Acquisition Reports*, Santa Monica, Calif.: RAND Corporation, MR-625-OSD, 1996. As of November 14, 2015:
http://www.rand.org/pubs/monograph_reports/MR625.html

Kerr, Julian, "Australian DoD Faces Sweeping Reforms," *Jane's Defence Weekly,* Vol. 48, No. 34, August 10, 2011.

———, "Australia Announces Major DoD Shakeup," *Jane's Defence Weekly,* Vol. 52, No. 20, April 1, 2015.

Klakegg, Ole Jonny, Terry Williams, Ole Morten Magnussen, and Helene Glasspool, "Governance Frameworks for Public Project Development and Estimation," *Project Management Journal,* Special Issue: Special PMI Research Conference, Vol. 39, No. S1, 2008, pp. S27–S42.

Klayman, Joshua, "Varieties of Confirmation Bias," in J. Busemeyer, R. Hastie, and D. L. Medin, eds., *Decision Making from a Cognitive Perspective,* New York: Academic Press, 1995, pp. 365–418.

Konkurrensverket, *2013 Annual Report,* Stockholm, 2013. As of November 11, 2015:
http://www.konkurrensverket.se/en/publications-and-decisions/annual-report-2013/

———, *2014 Annual Report,* Stockholm, 2014. As of November 11, 2015:
http://www.konkurrensverket.se/en/publications-and-decisions/annual-report-20141/

Kotter, John P., *Leading Change,* Boston, Mass.: Harvard Business School Press, 1996.

Kurian, George Thomas, ed., *The Encyclopedia of Political Science,* Washington, D.C.: CQ Press, 2011.

Leseure, Michel, "Risk," in *Key Concepts in Operations Management, Sage Key Concepts,* Thousand Oaks, Calif.: SAGE Publications, 2010, pp. 57–59.

Lewin, Kurt, *Frontiers in Group Dynamics,* The Bobbs-Merrill Reprint Series in the Social Sciences, S170, n.p., 1947.

Lo, Bernard, and Marilyn J. Field, eds., *Conflict of Interest in Medical Research, Education and Practice,* Washington, D.C.: Institute of Medicine of the National Academies, 2009.

Maravic, Patrick von, "Auditing," in Bertrand Badie, Dirk Berg-Schlosser, and Leonardo Morlino, eds., *International Encyclopedia of Political Science,* Thousand Oaks, Calif.: SAGE Publications, 2011, pp. 103–108.

Mayer, Lauren A., Mark V. Arena, and Michael McMahon, *An Excel Tool to Assess Acquisition Program Risk,* Santa Monica, Calif.: RAND Corporation, TL-113-OSD, 2013. As of November 14, 2015:
http://www.rand.org/pubs/tools/TL113.html

McGrath, Rita Gunther, "Exploratory Learning, Innovative Capacity and Managerial Oversight," *The Academy of Management Journal,* Vol. 44, No. 1, February 2001, pp. 118–131.

McRoberts, Flynn, "The Fall of Andersen," Part 1, *Chicago Tribune*, September 1, 2002.

Michaely, Roni, and Kent L. Womack, "Conflict of Interest and the Credibility of Underwriter Analyst Recommendations," *The Review of Financial Studies*, Vol. 12, No. 4, 1999, pp. 653–686.

MITRE Corporation, *Systems Engineering Guide*, McLean, Va., 2014.

Morse, Amyas, *Strategic Financial Management in the Ministry of Defence*, London: UK National Audit Office, 2015–2016.

NAO—*See* UK National Audit Office.

NASA, "Office of the Chief Financial Officer," web page, undated. As of September 17, 2015:
https://www.nasa.gov/offices/ocfo/home/#.VkQ8BtAmWxI

Naval Air Warfare Center, Training System Division, "Technical Reviews," web page April 8, 2015. As of November 14, 2015:
http://www.navair.navy.mil/nawctsd/Resources/Library/Acqguide/reviews.htm

Ndunguru, Cheryl, *Executive Core Qualifications: Becoming an Effective Leader, Senior Executive Resources and Performance Management, Training and Executive Development*, Washington, D.C.: U.S. Office of Personnel Management, undated.

Netherlands Ministry of Defence, *Overview of the Defence Materiel Process*, The Hague, September 2007.

———, *Doing Business with the Netherlands Ministry of Defence*, The Hague, 2008.

Neubaum, Donald, "Stewardship Theory," in Eric H. Kessler, ed., *Encyclopedia of Management Theory*, Thousand Oaks, Calif.: SAGE Publications, 2013, pp. 768–770.

New Zealand Defence Force, *Briefing for the Incoming Minister of Defence*, Wellington: Ministry of Defence, October 2014. As of November 12, 2015:
http://www.nzdf.mil.nz/corporate-documents/briefing-for-incoming-minister.htm

———, *The 2014–2015 Annual Report: For the Year Ended 30 June 2015*, Wellington: Ministry of Defence, 2015. As of November 16, 2015:
http://www.nzdf.mil.nz/downloads/pdf/public-docs/nzdf-annual-report-2015.pdf

Nickerson, Raymond S., "Confirmation Bias: A Ubiquitous Phenomenon in Many Guises," *Review of General Psychology*, Vol. 2, No. 2, 1998, pp. 175–220.

Nieminen, Levi, and Daniel Denison, "Organisational Culture and Effectiveness," in Eric H. Kessler, ed., *Encyclopedia of Management Theory*, Thousand Oaks, Calif.: SAGE Publications, 2013, pp. 530–531.

Oleszek, Walter J., *Congressional Oversight: An Overview*, Washington, D.C.: Congressional Research Service, February 22, 2010. As of October 7, 2015:
http://fas.org/sgp/crs/misc/R41079.pdf

O'Reilly, C. A., III, and J. A. Chatman, "Culture as Social Control: Corporations, Cults, and Commitment," *Research in Organizational Behavior*, Vol. 18, 1996, pp. 157–200.

"Organisation," web page, Bundeskartellamt.de, undated. As of November 16, 2015:
http://www.bundeskartellamt.de/EN/AboutUs/Bundeskartellamt/Organisation/organisation_node.html

Overview and Scrutiny Commission, *Overview and Scrutiny at the London Borough of Merton*, London: London Borough of Merton, March 2006.

Public Law 108-458, Intelligence Reform and Terrorism Prevention Act of 2004, Section 1017, Alternative Analysis of Intelligence by the Intelligence Community, December 17, 2004.

Reed, B. J., and John W. Swain, *Public Finance Administration*, 2nd ed., Thousand Oaks, Calif.: SAGE Publications, 1997.

Ryan, Richard M., and Edward L. Deci, "Intrinsic and Extrinsic Motivations: Classic Definitions and New Directions," *Contemporary Educational Psychology*, Vol. 25, 2000, pp. 54–67.

Sevic, Zeljko, "Auditing Standards," in Charles Wankel, ed., *Encyclopedia of Business in Today's World*, Thousand Oaks, Calif.: SAGE Publications, 2009, pp. 91–93.

Simon, Herbert A., *Administrative Behavior: A Study of Decision-Making Processes in Administrative Organizations*, 4th ed., New York: The Free Press. 1997.

Smith, Robert, *Audit Committees Combined Code Guidance: A Report and Proposed Guidance by an FRC-Appointed Group Chaired by Sir Robert Smith*, London: Financial Reporting Council, January 2003.

Sprague, Robert, and Sean Valentine, "Due Diligence," in Robert W. Kolb, ed., *Encyclopedia of Business Ethics and Society*, Thousand Oaks, Calif.: SAGE Publications, 2008, pp. 624–628.

Staronova, Katarina, "Regulatory Impact Assessment: Formal Institutionalization and Practice," *Journal of Public Policy*, Vol. 30, No. 1, 2010, pp. 117–136.

Stockholm International Peace Research Institute, SIPRI Military Expenditure Database, undated. As of October 21, 2015:
http://www.sipri.org/research/armaments/milex/milex_database

Sturgess, Gary L., *Contestability in Public Services: An Alternative to Outsourcing*, ANZSOG Research Monograph, Melbourne: Australia and New Zealand School of Government, April 2015. As of October 8, 2015:
https://www.anzsog.edu.au/media/upload/publication/150_Sturgess-Contestability-in-Public-Services.pdf

Sutcliffe, Kathleen, "Organisational Culture Model," in Eric H. Kessler, ed., *Encyclopedia of Management Theory*, Thousand Oaks, Calif.: SAGE Publications, 2013, pp. 531–536.

Swedish Defence Research Agency, "Long Term Defence Planning—Purpose and Context," briefing, Stockholm, November 12, 2012.

Taubman, Peter M., "Audit Culture," in Craig Kridel, ed., *Encyclopedia of Curriculum Studies*, Thousand Oaks, Calif.: SAGE Publications, 2010, pp. 60–62.

UK Ministry of Defence, *Defence Audit Committee: Terms of Reference*, London, March 2013.

———, *How Defence Works*, Version 4.1, London, September 30, 2014.

UK MoD—*See* UK Ministry of Defence.

UK National Audit Office, home page, undated. As of September 18, 2015:
http://www.nao.org.uk

United States Code, Title 10, Section 2334, Independent Cost Estimation and Cost Analysis, December 26, 2013.

United States Code, Title 50, Section 2659, Report on Security Vulnerabilities of National Laboratory Computers, 2011.

U.S. Department of Defense Directive 7045.14, *The Planning, Programming, Budgeting, and Execution (PPBE) Process*, Washington, D.C.: Under Secretary of Defense (Comptroller), January 25, 2013.

U.S. Department of Defense Instruction 5000.02, *Operation of the Defense Acquisition System*, Washington, D.C.: Under Secretary of Defense for Acquisition, Technology, and Logistics, January 7, 2015.

U.S. Government Accountability Office, "About GAO," web page, undated-a. As of October 7, 2015:
http://www.gao.gov/about/index.html

———, home page, undated-b. As of September 18, 2015:
http://www.gao.gov

———, "Our Workforce," web page, undated-c. As of October 7, 2015:
http://www.gao.gov/about/workforce/

———, "Performance Measures," web page, undated-d. As of October 8, 2015:
http://www.gao.gov/about/perfmeasures.html

———, *Case Study Evaluations*, Washington, D.C., GAO/PEMD-91-10.1.9, November 1990.

———, *Best Practices: Better Management of Technology Development Can Improve Weapon System Outcomes*, Washington, D.C., GAO/NSIAD-99-162, July 1999.

————, *Weapon System Acquisitions: Opportunities Exist to Improve the Department of Defense's Portfolio Management*, Washington, D.C., GAO-15-466, August 2015. As of October 13, 2015:
http://www.gao.gov/assets/680/672205.pdf

U.S. Office of Cost Assessment and Program Evaluation, "About CAPE," undated. As of September 17, 2015:
http://www.cape.osd.mil

————, *Operating and Support Cost-Estimating Guide*, Washington, D.C., March 2014.

U.S. Securities and Exchange Commission, Office of Compliance Inspections and Examinations, "Investment Adviser Due Diligence Processes for Selecting Alternative Investments and Their Respective Managers," *National Exam Program Risk Alert*, Vol. 4, No. 1, January 28, 2014. As of October 7, 2015:
http://www.sec.gov/about/offices/ocie/
adviser-due-diligence-alternative-investments.pdf

Værnsfælles Forsvarskommando, "The Danish Defence Acquisition and Logistics Organization," web page, undated. As of November 11, 2015:
http://forsvaret.dk/fmi/eng/Pages/default.aspx

Wagner-Tsukmoto, Sigmund, "Scientific Management," in Eric H. Kessler, ed., *Encyclopedia of Management Theory*, Thousand Oaks, Calif.: SAGE Publications, 2013, p. 679.

World Bank, *Best Practices for Internal Audit in Government Departments*, undated. As of November 4, 2015:
http://siteresources.worldbank.org/EXTFINANCIALMGMT/Resources/
313217-1196229169083/4441154-1196269165212/4443896-1196270435064/
BestPracIntAuditGovDepts.pdf